Biodiversity

CONSERVING ENDANGERED SPECIES

GREEN TECHNOLOGY

Biodiversity

CONSERVING ENDANGERED SPECIES

Anne Maczulak, Ph.D.

Facts On File
An imprint of Infobase Publishing

BIODIVERSITY: Conserving Endangered Species

Copyright © 2010 by Anne Maczulak, Ph.D.

Facts On File, Inc.
An imprint of Infobase Publishing
132 West 31st Street
New York NY 10001

Library of Congress Cataloging-in-Publication Data
Maczulak, Anne E. (Anne Elizabeth), 1954–
   Biodiversity : conserving endangered species / author, Anne Maczulak.
     p. cm.—(Green technology)
   Includes bibliographical references and index.
   ISBN-13: 978-0-8160-7197-5
   ISBN-10: 0-8160-7197-7
   1. Biodiversity—Juvenile literature. 2. Endangered species—Juvenile literature.
I. Title.
   QH541.15.B56M33 2010
   639.9—dc22   2008051758

Facts On File books are available at special discounts when purchased in bulk quantities for businesses, associations, institutions, or sales promotions. Please call our Special Sales Department in New York at (212) 967-8800 or (800) 322-8755.

You can find Facts On File on the World Wide Web at http://www.factsonfile.com

Text design by James Scotto-Lavino
Illustrations by Bobbi McCutcheon
Photo research by Elizabeth H. Oakes

Printed in the United States of America

Bang Hermitage 10 9 8 7 6 5 4 3 2

This book is printed on acid-free paper.

Contents

Preface

The first Earth Day took place on April 22, 1970, and occurred mainly because a handful of farsighted people understood the damage being inflicted daily on the environment. They understood also that natural resources do not last forever. An increasing rate of environmental disasters, hazardous waste spills, and wholesale destruction of forests, clean water, and other resources convinced Earth Day's founders that saving the environment would require a determined effort from scientists and nonscientists alike. Environmental science thus traces its birth to the early 1970s.

Environmental scientists at first had a hard time convincing the world of oncoming calamity. Small daily changes to the environment are more difficult to see than single explosive events. As it happened the environment was being assaulted by both small damages and huge disasters. The public and its leaders could not ignore festering waste dumps, illnesses caused by pollution, or stretches of land no longer able to sustain life. Environmental laws began to take shape in the decade following the first Earth Day. With them, environmental science grew from a curiosity to a specialty taught in hundreds of universities.

The condition of the environment is constantly changing, but almost all scientists now agree it is not changing for the good. They agree on one other thing as well: Human activities are the major reason for the incredible harm dealt to the environment in the last 100 years. Some of these changes cannot be reversed. Environmental scientists therefore split their energies in addressing three aspects of ecology: cleaning up the damage already done to Earth, changing current uses of natural resources, and developing new technologies to conserve Earth's remaining natural resources. These objectives are part of the green movement. When new technologies are invented to fulfill the objectives, they can collectively be called green technology. Green Technology is a multivolume set that explores new methods for repairing and restoring the environment. The set covers a broad range of subjects as indicated by the following titles of each book:

- *Cleaning Up the Environment*
- *Waste Treatment*
- *Biodiversity*
- *Conservation*
- *Pollution*
- *Sustainability*
- *Environmental Engineering*
- *Renewable Energy*

Each volume gives brief historical background on the subject and current technologies. New technologies in environmental science are the focus of the remainder of each volume. Some green technologies are more theoretical than real, and their use is far in the future. Other green technologies have moved into the mainstream of life in this country. Recycling, alternative energies, energy buildings, and biotechnology are examples of green technologies in use today.

This set of books does not ignore the importance of local efforts by ordinary citizens to preserve the environment. It explains also the role played by large international organizations in getting different countries and cultures to find common ground for using natural resources. *Green Technology* is therefore part science and part social study. As a biologist, I am encouraged by the innovative science that is directed toward rescuing the environment from further damage. One goal of this set is to explain the scientific opportunities available for students in environmental studies. I am also encouraged by the dedication of environmental organizations, but I recognize the challenges that must still be overcome to halt further destruction of the environment. Readers of this book will also identify many challenges of technology and within society for preserving Earth. Perhaps this book will give students inspiration to put their unique talents toward cleaning up the environment.

Acknowledgments

I would like to thank a group of people who made this book possible. Appreciation goes to Bobbi McCutcheon, who helped turn my unrefined and theoretical ideas into clear, straightforward illustrations, and to photo editor Elizabeth Oakes, for her wonderful contributions. Special gratitude is due to Melanie Piazza, director of animal care at WildCare in San Rafael, California, for providing details on wildlife rehabilitation and release. My thanks also go to Marilyn Makepeace, who provided support and balance to my writing life, and Jodie Rhodes, who helped me overcome more than one challenge. Finally, I thank Frank Darmstadt, Executive Editor, for his patience and encouragement throughout my early and late struggles to produce a worthy product. General thanks go to Facts On File for giving me this opportunity.

Introduction

Students with a modest understanding of the environment and concern for its future know it is important to preserve biological diversity. *Biodiversity* is the variety of living things on Earth or in a specific area. This definition seems simple enough to understand, yet the concept of biodiversity has deeper meanings that challenge even trained environmental scientists. Defining biodiversity becomes not unlike attempts to find the essence of peace, wealth, or happiness; they mean different things to different people.

A region that has a wide variety of *species* in robust *populations* is said to possess biodiversity. But not every place on Earth bursts with diverse life. This does not mean the Earth's biodiversity is gone. Biodiversity concentrates in certain areas, while other parts of the globe possess a somewhat lesser variety and number of species. Healthy *ecosystems* require larger population sizes of certain species such as plants and small prey animals to serve as food for other animals, and some other species must have small population sizes to reduce competition between individuals. For example, in a mountain ecosystem, predators such as mountain lions exist in much smaller numbers than deer, which serve as mountain lion prey, yet insects live in huge numbers because they are food for a variety of species of other insects, birds, mammals, reptiles, amphibians, and fish. Environmental scientists must understand this normal variability in nature to assess biodiversity and species loss.

Scientists delving deeper into the concept of biodiversity have pointed out that biodiversity also means the variability of species' total genes, also known as genetic biodiversity. On a large scale, the environment requires diverse ecosystems to carry out cycling of energy, matter, and nutrients. This is ecological biodiversity. Biodiversity loss has grown to crisis levels in many parts of the world, and these losses are caused by genetic, species, and ecological biodiversity loss. Of all threats to endangered plants and animals, *habitat* loss is the greatest.

Two important aspects of biodiversity are habitat and *adaptation*. Each of these concepts contributes to the overall theme of this book because they explain how biodiversity grows and how it can decline. Habitat is where an organism lives. This book examines the many ways in which habitat is destroyed and how this has a devastating effect on biodiversity. Adaptation is an organism's acquiring a trait that helps it survive. A species' ability or inability to adapt to changes in its environment has contributed to the evolution of biodiversity on Earth. It also helps explain the ways in which biodiversity disappears in the face of human population growth.

To save biodiversity, environmental scientists use one of two approaches. Each appreciates the importance of both habitat and adaptation, but each approach stresses them in different ways. The ecosystem approach to saving biodiversity focuses on total habitats to protect the ecosystems within them; this approach saves species by ensuring healthy ecosystems for those species. This volume examines the second approach, the species approach, which focuses primarily on individual species to save biodiversity. These two approaches to solving biodiversity loss are not in conflict, and this book points out where they overlap.

The first chapter reviews the issue of biodiversity, what it means and why it can cause controversy. It also covers various influences on biodiversity, such as climate change and environmental ethics. The second chapter provides more detail on these subjects. It describes the types of biodiversity, the different roles of species in ecosystems, *extinction,* and the thorniest problem of all: estimating the actual number of species on Earth.

Chapters 3 and 4 cover two important threats to biodiversity. These are *invasive species* and urban development. These chapters explain how each threatens ecosystems, which in turn places undue stresses on species. In its discussion of invasion, chapter 3 covers the basics of adaptation. It also includes a description of methods for eradicating invasive species. Chapter 4 covers the various hazards for animals from urbanization. It discusses the means by which urban planners can now balance urbanization with the needs of wildlife.

Chapter 5 covers nature reserves. It introduces the basics of wildlife conservation and specialized ways for preserving *threatened species.* Zoos, aquariums, sanctuaries, and artificial habitats cause their share of debate, but they may become the last hope for a number of critically endangered animals.

In chapter 6, various topics related to saving biodiversity are discussed. It presents overviews of global conservation strategies, methods for saving marine biodiversity, the idiosyncrasies of monitoring wildlife, and the process of removing a species from the *endangered species* list.

Chapter 7 details the methods by which scientists study the topics presented in this book. It describes how to measure two important components of biodiversity: *richness* and *rarity*. The chapter covers the monitoring of plants, animals, amphibians and reptiles, and aquatic species. This chapter also concludes with a summary of the criteria for deciding whether a species is endangered or abundant.

Overall, this book offers a review of biodiversity technology. Biodiversity also covers the many ways in which people interfere with ecosystems and habitats, as well as the ways scientists restore them. The book also explores the ethical questions that arise when trying to rescue threatened species in the face of dire human conditions. Biodiversity loss is truly a multifaceted crisis. It is caused mainly by human activities, but as the rate of species loss increases, biodiversity loss will also impact the well-being of people throughout the world.

ENDANGERED SPECIES

Endangered species are any wildlife or plants with so few individuals in their population they could soon become extinct in their natural range. Biodiversity and the quantity of endangered species on Earth are therefore intertwined. Endangered status may lead to extinction, and extinction results in biodiversity loss, so in a sense endangered species also symbolize the success or failure of today's scientific attempts to save biodiversity.

The Endangered Species Act, passed by the U.S. Congress in 1973, provides protections for species in the United States. In 1966, 78 species had received formal protection under the Endangered Species Preservation Act, but the stronger Endangered Species Act marked a milestone by being the first legislation to take the unprecedented step of giving legal protection to species other than humans. Upon signing the law, President Richard M. Nixon stated, "Nothing is more priceless and more worthy of preservation than the rich array of animal life with which our country has been blessed. It is a many-faceted treasure, of value to scholars, scientists, and nature lovers alike, and it forms a vital part of the heritage we all share as Americans." This idea, however, has led to considerable controversy because it introduced a new concept to the public's way of thinking: All species, not just humans, have a right to live and thrive on this planet. The Endangered Species Act's rocky history has in fact come about after two centuries of philosophical differences on the role of nature to society.

In the 1800s hunters in the Midwest and Southeast saw a steady decrease in the numbers of waterfowl and shorebirds due to commercial hunting for plumes and feathers to be sold to clothiers in eastern cities. Game hunters urged their leaders, including Iowa Congressman John Lacey, to help stem the loss of herons, egrets, and parakeets. With the hope of preventing outsiders from depleting Iowa's hunting stock for the state's own sportsmen, Lacey drafted a bill that prevented hunters from crossing state lines to kill game birds—Iowa's hunters certainly sold their share of game, much of which ended up in fur coats and feathers for women's hats. Lacey and the hunters he represented convinced Washington of the need for wildlife protection, and President William McKinley signed the Lacey Act into law in 1900. Soon after its passage, Lacey addressed a meeting of women's groups in Iowa with a lighthearted admonishment: "In the preservation of our birds, the women of America were slow to act, but they are now doing a great part. We have a wireless telegraph, a crownless queen, a thornless cactus, a seedless orange, and a coreless apple. Let us now have a birdless hat!" The Lacey Act led the way toward a new era of wildlife protections.

Poachers almost immediately set about circumventing the new law, so Congress reacted with additional species protections. President Theodore Roosevelt took the most symbolic and significant step in wildlife protection in 1903 by creating the first National Wildlife Refuge in Florida. The new reserve would provide the most hunted shorebirds with a safe refuge for breeding. Thirteen years later Roosevelt reemphasized his thoughts on preservation in his *A Book-Lover's Holidays in the Open*: "Defenders of the short-sighted men who in their greed and selfishness will, if permitted, rob our country of half its charm by their reckless extermination of all useful and beautiful wild things sometimes seek to champion them by saying the 'game belongs to the people.' So it does; and not merely to the people now alive, but to the unborn people." This began what would become a long career in the preservation of the American wilderness.

Despite provisions for creating refuges, the United States continued to lose species. The 1913 Weeks-McLean Law helped a little by outlawing the hunting of migratory birds for commercial purposes, and the more comprehensive Migratory Bird Treaty Act of 1918 targeted specific species, and may well have saved plume birds, such as the snowy egret, even while others disappeared as the 20th century dawned. The Migratory Bird Treaty Act also became the first act to originate from international talks rather than local or national efforts. The United States, Canada, Mexico, Japan, and Russia had discussed the need to protect bird migratory routes that these countries had in common. Logically, protections enacted in one country but no other would have little effect on helping birds that traveled halfway around the globe each year.

Early attempts at protecting wildlife resulted from the desire for beautiful plumage from some of the world's most showy and colorful birds. Most animals populating the United States in the early 1900s, however, did not possess spectacular features. One obscure animal, little known to the public, introduced the challenges and questions that would arise for devising the best approaches in wildlife protection. Scientists had known since the 1890s of an inch-long fish found only in steaming geothermal labyrinths beneath the Mojave Desert in California's Death Valley National Park. Biologists had named it the Devil's Hole pupfish after one of the desert sites where it lived. Curious biologists descended upon the sun-baked Mojave to glimpse the rare creature, perhaps one of the rarest animals in the world. In the 1940s pupfish numbers had begun a steady decline as urban centers sprawled west and thirsty communities pumped water from the aquifers beneath the desert. By the 1950s the fish's total population had plummeted to fewer than 200 individuals. In 1952 the scientific community encouraged President Harry S. Truman to place protections around not only the pupfish but also its habitat. Mike Bower, a National Park Service fish biologist, explained to the *San Francisco Chronicle* in 2007, "This fish is the species that made us take note of our need for conservation. It made us realize that our actions have an impact beyond us." Habitat protection became a new step forward in federal wildlife protection, and the pupfish became the first animal to receive habitat protection.

In 1966 Congress sought to provide further safeguards for the needs of duck hunters and fishermen by passing the Endangered Species Preservation Act, which allowed the U.S. Fish and Wildlife Service (FWS) to acquire land for conserving "selected species of native fish and wildlife" that the government deemed were the most important to the hunting and fishing industries. The FWS did not have a clear plan for setting aside land, and many wildlife populations continued to dwindle, but at least the act provided a means for listing the status of individual species. By the end of the 1960s species decline had become quite evident, so Congress began to close loopholes in the Preservation Act that had allowed these declines: military bases, businesses, and landowners that continued to hunt without restraint and destroy habitat. In 1973 President Nixon signed into law the Endangered Species Act. The new, stronger law addressed wildlife worldwide and added mammals, amphibians, invertebrates and crustaceans, and plants to the list of waterfowl and fish already covered. The act made killing, harming, or other forms of "take" illegal, meaning a person could not remove a listed animal from its natural habitat for any reason. Most important, species along with habitat would receive protection.

Today the FWS and the National Marine Fisheries Service jointly enforce the Endangered Species Act, which distinguishes among endangered species, those at most serious risk of becoming extinct, and threatened, species currently abundant in their natural range but likely to become endangered because of declining numbers. The two agencies oversee a daily updated list that contains about 1,100 animal species and 750 plant species.

Opponents of wildlife protections, including this hunter quoted in *Audubon* magazine in 2007, have argued that people should be allowed to kill endangered or threatened species, in this case, wolves living in Yellowstone National Park: "Who wants to bring their family up here and camp out and worry about their kids being taken down by a wolf?" A dramatic example to be sure, but at the very least, landowners want compensation

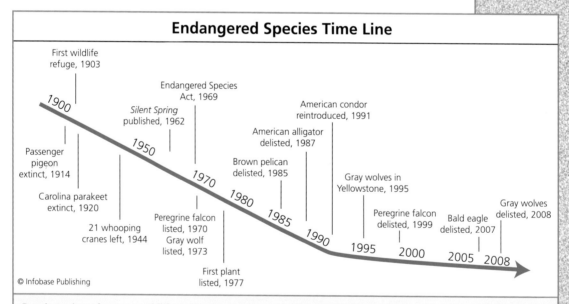

Endangered Species Time Line

First wildlife refuge, 1903

Endangered Species Act, 1969

Silent Spring published, 1962

American condor reintroduced, 1991

American alligator delisted, 1987

Brown pelican delisted, 1985

Gray wolves in Yellowstone, 1995

Gray wolves delisted, 2008

1900

1950

1970

1980

1985

1990

1995

2000

2005   2008

Passenger pigeon extinct, 1914

Carolina parakeet extinct, 1920

21 whooping cranes left, 1944

Peregrine falcon listed, 1970

Gray wolf listed, 1973

First plant listed, 1977

Peregrine falcon delisted, 1999

Bald eagle delisted, 2007

© Infobase Publishing

Despite serious threats to wildlife in the past, species-protection laws and conservation groups have made progress in reestablishing species that were once endangered. The return of breeding populations of gray wolves to the northern Rocky Mountains marks a recent and heavily publicized success.

when the law says they must protect habitat located on their land. Others argue that the most seriously endangered species should be allowed to disappear so that biologists can concentrate on species easier to save. In many cases, opponents have reasoned, some species come and go without humans having any knowledge of them. Terry Baldino of Death Valley National Park has countered, "We need to tell people they're [pupfish] as valuable as the bald eagle and worth saving. It'd be real scary for the fish to disappear and never know it. That'd be a tragedy because there's so much we could learn from it. You have to look at the bigger picture. This is just the tip of the iceberg. The aquifer doesn't even originate in this part of Nevada. . . . This species is like a canary in a coal mine. If this species goes, it's a trigger. What else goes?" Both opponents and supporters of wildlife protections have considered the big picture as Baldino suggested, but they have come to diametrically different views.

This chapter examines the biodiversity debate and how it differs for familiar species compared with obscure species. The chapter also discusses the multifaceted question of why biodiversity is important. It covers the mechanisms of wildlife protection, climate change's effect on biodiversity, and conservation biology, which focuses on ways to preserve species. It also delves into the challenges and the debates that surround biodiversity loss. Finally, the chapter introduces aspects of environmental ethics.

THE BIODIVERSITY DEBATE

Sometimes the needs or desires of people conflict with the things an endangered species requires for survival. When this occurs, a debate about biodiversity unfolds, which at its core pits the value of animal and plant life against the value of human life. The Endangered Species Act has a mandate to protect threatened habitats as well as species, but when habitat lies on private land, many landowners feel the law intervenes with their right to use their property as they see fit. Utah State University political science professor Randy T. Simmons has opined, "Under the current law, landowners are punished for owning habitat that attracts or protects an endangered species. This fact leads rational, normally law-abiding citizens to destroy habitat before an endangered species arrives." Mr. Simmons's theory has indeed played out in Boiling Spring Lakes, North Carolina, where property owners have cleared their land of pines that are habitat to the endangered red-cockaded woodpecker. Landowner Steve Lowery owned two vacant lots and decided to level one of them of all its trees. "I get one lot and the woodpeckers get the other one," he told Associated Press in 2006. Of course, protecting habitat takes more than the simple compromise Lowery proposed, partially in jest. When an endangered species means a tiny amphibian or a delicate lichen, protections become even more difficult than for a bird or mammal. Often businesses and local governments pressure the federal government to weaken the law to preserve their income. This scenario has occurred several times since the enactment of the Endangered Species Act.

Biodiversity must be discussed in terms of value in order for many people to understand what it means to lose it. Environmentalists therefore often speak of biodiversity as a benefit. As yet undiscovered species may in the future serve as sources for new medicines, food, cosmetics, or enzymes for cleaning up pollution and recycling nutrients. Today, however, many—but certainly not all—new drugs come out of laboratories rather than a jungle, and genetic engineering provides more and more products for consumers and for environmental cleanup. These advances may lead some to question the need for biodiversity in the modern world.

The following three main themes occur in the biodiversity debate:

1. How extensive are the current threats to biodiversity?

2. How can these proposed threats be measured in a reliable manner?

3. Do humans really have a role in protecting biodiversity?

Environmental ethicists have proposed two ways to view the value of Earth's biodiversity: *instrumental value* and *intrinsic value*. The instrumental value of biodiversity proposes that the Earth is a resource to be used as humans see fit, for deriving new medicines from exotic plants, new antibiotics from animals or microbes, pesticides, unique food sources, or industrial materials. After human needs are met, people may be more willing to help other species survive. Instrumental value consists of two subcategories: use value and nonuse value. Use values benefit people in economic, scientific, or recreational ways, and people may therefore be willing to protect them if they feel a benefit may someday come from these resources. Nonuse values, by contrast, give benefits even when people seldom come in contact with the item. For example, nonuse value resides in the knowledge that an old-growth forest is home to spotted owls, even if that forest is hundreds of miles away. Other examples of nonuse value are provided by forests, marshes, mountain ranges, and jungles because of their solitude, quiet, or natural beauty.

Intrinsic value of biodiversity suggests that the Earth exists not solely for human use but serves a role in the universe beyond human needs. Intrinsic value does not translate into direct benefits for people but, rather, relates to a person's beliefs about the meaning of life. Civilization has confronted this question since the beginning of recorded history, so it is not surprising that the intrinsic value of nature can be difficult to define.

The northern spotted owl of the U.S. Pacific Northwest provides an example of how views differ on biodiversity, as well as the instrumental versus intrin-

The nocturnal bird of prey, the northern spotted owl, inhabits old-growth forests in the Pacific Northwest and northern California. It became a symbol of the value and the frustrations of species protection because its threatened status was designated based on the bird's habitat and not on the total number of spotted owls living in that habitat. (*John and Karen Hollingsworth, U.S. Fish and Wildlife Service*)

sic value of nature. This nocturnal hunter feeds on flying squirrels and nests only in the cavities at the top of broken old-growth fir trees. This is a very specialized creature! Old-growth forests contain large, mature trees that provide the spotted owl's favorite habitat, but in the past 200 years the timber industry has reduced old-growth forests in the United States by 95 percent. As the trees fell, the spotted owl struggled to find habitat; the government added it to the endangered species list in 1990 (as threatened) and listed its habitat in 1992. The protections to the northern spotted owl created a fiery debate over nature's instrumental value and intrinsic value.

Timber jobs in the Pacific Northwest began to vanish and with them, the land's instrumental value. The spotted owl incident led, nevertheless, to new techniques in forest management. Clear-cutting, which results in splitting habitat into separate pieces, has stopped. A government-initiated program called the Forest Plan coordinates activities between the Bureau of Land Management (BLM), the U.S. Forest Service, and the FWS to develop ways for maintaining habitat and retaining a lumber-based industry. The area has also welcomed tourists seeking to experience the intrinsic benefits of deep, old-growth forests.

Scientists and philosophers may never find a fail-safe way to measure biodiversity's value. Most likely, biodiversity's value comes from a spectrum of benefits from the tangible (money) to the intangible (beauty).

ENDANGERED AND THREATENED SPECIES

The first animal protection laws came about because people viewed wildlife as assets: food, trophies, or recreation. Today the Endangered Species Act and international laws aim to save species both for their use and non-use values and for their intrinsic values. The act protects threatened species to the same extent as it protects endangered species.

An endangered species is an animal or plant in immediate danger of becoming extinct throughout all or a significant portion of its range. A threatened species has sufficient numbers in its natural range at the moment, but its numbers are declining and it will likely become endangered in the near future. The three main protections provided by the Endangered Species Act are the following: protection against hunting, protection against removal of an animal from its habitat, and protection against the destruction of critical habitats, which are those believed to be essential for the survival of a species.

The FWS enforces the Endangered Species Act; as part of this service it maintains a list of all the U.S. endangered or threatened species. The following table shows the main plant and animal groups on the endangered species list. These groupings and particular species may change periodically as science reveals more about the genetic makeup of plants and animals. The classifications of living things are discussed further in the "Carl Linnaeus—The Father of Taxonomy" sidebar on page 23.

THE ENDANGERED SPECIES LIST CATEGORIES	
ANIMAL GROUPS	**PLANT GROUPS**
amphibians	conifers and cycads (leafy trees)
arachnids	ferns
birds	flowering plants
clams	lichens
corals	
crustaceans	
fishes	
insects	
mammals	
reptiles	
snails	

The International Union for Conservation of Nature (IUCN, also called the World Conservation Union) publishes the *Red List,* on the current status of the world's animals and plants. The Red List differs from the endangered species list because it includes all species, even those that have gone extinct (since 1500 B.C.E.), plus species that are abundant. Like the endangered species list, the Red List categories change as some populations become more threatened and other populations recover. The following "Case Study: The Ivory-Billed Woodpecker," on page 11, examines an instance of a species changing from abundant to rare to possible extinction.

THE WEB OF LIFE

Energy flows from the Sun to the Earth in a one-way direction. Green plants play a role in this flow as *producers* when they absorb the Sun's radiant energy and convert inorganic nutrients and carbon dioxide to organic compounds. Thus plants produce forms of energy and matter that can be used by animals. Those animals that eat plants for energy and nutrients are called *consumers,* specifically, they are primary consumers because they are the first step in energy exchange from plant life to animal life. Some

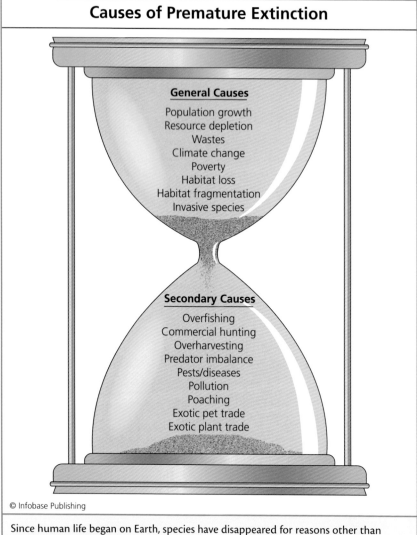

Causes of Premature Extinction

General Causes

Population growth
Resource depletion
Wastes
Climate change
Poverty
Habitat loss
Habitat fragmentation
Invasive species

Secondary Causes

Overfishing
Commercial hunting
Overharvesting
Predator imbalance
Pests/diseases
Pollution
Poaching
Exotic pet trade
Exotic plant trade

© Infobase Publishing

Since human life began on Earth, species have disappeared for reasons other than natural causes. In the Earth's history, however, many species, such as dinosaurs, have become extinct owing to natural causes.

consumers feed on living prey—a large fish ingesting a smaller fish—and others feed on matter that is no longer living—a human eating a steak dinner. These animals represent secondary, tertiary, or quaternary consumers. In biology most consumers are also called heterotrophs, organisms that depend on a wide variety of substances for their nutrition.

CASE STUDY: THE IVORY-BILLED WOODPECKER

The large ivory-billed woodpecker once nested throughout the southeastern United States and thrived in great expanses of virgin woodland that covered much of the region before the Civil War. These tracts of swampland, nicknamed the "Big Woods," surrounding Arkansas's Cache River, are called bottomland hardwoods and contain many dead and dying trees. Bottomland hardwoods attract beetles and so become a source of beetle larvae, the ivory-bill's favorite food. After the Civil War, the lumber industry cut down swaths of hardwoods for building homes, and by the 1940s the bottomland had shrunk. Few residents living nearby saw the woodpecker again. On March 11, 1967, the FWS added the ivory-billed woodpecker to the endangered species list, though most biologists felt it was already extinct.

In 2004 a current of excitement ran through the world of ornithology. A kayaker paddling the Cache River had spotted a bird he believed was an ivory-billed woodpecker. Soon afterward two others caught on film a brief glimpse of a bird having the woodpecker's characteristic markings. The director of the Cornell Lab of Ornithology, John Fitzpatrick, told *National Geographic*, "Through the 20th century it's been every birder's fantasy to catch a glimpse of this bird, however remote the possibility. This really is the holy grail." Recordings taken in the densest parts of the bottomlands gave evidence of the ivory-bill's distinctive double rap-rap on tree trunks. Birdwatchers converged on the Cache River and nearby White River National Wildlife Refuge, the last remaining places believed to support the special bird. Frank Gill of New York's Audubon Society remarked, "It is kind of like finding Elvis." Since 2004 persistent volunteers have made no additional sightings, though the Cornell group and the Nature Conservancy have devoted 3,000 hours of in-person searches and movement-controlled photography. Robotic sensors scan the woods for any and all movements, while the Automated Collaborative Observatory for Natural Environments (ACONE) searches the Arkansas skies on the lookout for birds of any type.

Did a species thought to be extinct somehow manage to recover and begin to repopulate the area? The last tentative sighting occurred in 2005,

(continues)

(continued)

but ecologists have wisely gone on the offensive in preserving what is left of the woodpecker's habitat. Mr. Fitzpatrick told the *Boston Globe* in 2008, "The decline of the ivory-billed is an unspeakable American tragedy. This country was unable to save even a single square meter of pristine bottom-land habitat. It all went under the ax and chainsaw. We may have lost this iconic bird, but, by God, we owe the ivory-billed this sort of exhaustive, scientific search. . . . If they are there, we also owe them a recovery program." The Nature Conservancy has worked jointly with the FWS to set aside thousands of acres of woodpecker habitat. "The successful history of conservation in the Big Woods of Arkansas," said the Nature Conservancy's Scott Simon, "is the result of great partnerships—federal and state agencies working with other organizations, local communities, hunters and landowners." The emotional support from people like Fitzpatrick plus government support provide the best chance for saving critically endangered species.

In 2007 the FWS designed a recovery plan for the ivory-billed woodpecker, should it indeed still hide in the swamps. The agency focuses on the following three goals: verify the existence of the bird by sightings, recordings, or nest cavities in trees; protect or add to current habitat; and study all factors that would threaten a potential ivory-billed colony. In an encouraging example of positive thinking, the FWS has set a goal of 2075 for the year in which the ivory-billed woodpecker will be removed from the endangered species list, a step called *delisting*.

The scientific community's and the public's reaction to the ivory-billed woodpecker sighting attests to the current state of many species that were once abundant. The ivory-billed woodpecker experience—whether or not the bird still lives—has offered good examples of the quick reactions and sound planning needed to preserve disappearing species.

To conserve the Earth's nutrients and make them available to other species, the Earth relies on organisms called *decomposers*. Decomposers are living things, usually microbes, that break down matter so it can be recycled. Radiant energy from the Sun does not recycle and is lost, but due to the action of decomposers nutrients cycle continuously through soil, water, and the atmosphere. These producer-consumer-decomposer relationships collectively make up ecosystems. On a larger scale, the many interrelationships between nonliving matter and living things make up what is known as a *web of life* or a food web.

Diversity among the members of a food web contributes to ecosystem survival. Food webs are complex networks of interconnected food chains. These webs depend on a diverse collection of *biota*, a term for living things,

Ecosystem Energy and Matter Flow

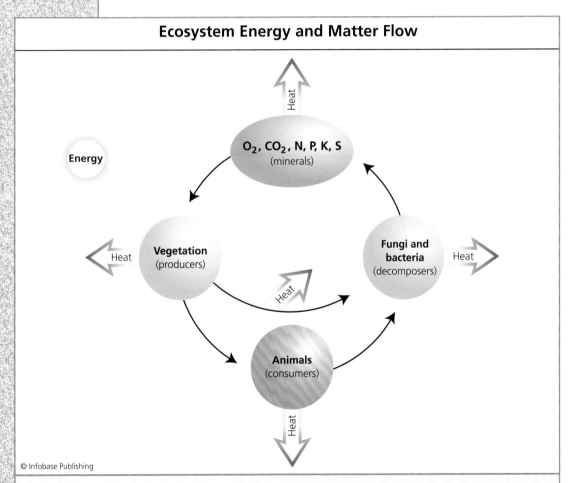

© Infobase Publishing

All plant life, animal life, and microorganisms contribute to the continual recycling of the Earth's elements. According to the second law of thermodynamics, energy as heat is lost with each step in the cycle. Organisms that perform photosynthesis capture energy from the Sun and so replenish the energy-matter cycle.

to make the web stable and resilient. (A food chain is a series of organisms, usually increasing in size, in which each one eats the preceding, or smaller, member.) Species diversity in a food web helps make the web resilient in case part of the web becomes damaged. If one nutrient-energy route were to be shut down, by pollution, for instance, other routes continue to provide nutrients and energy all the way to the top of the web's food chains. By contrast simple food webs make life riskier for their members because simple webs leave each animal up the food chain more and more vulnerable should anything go wrong.

Energy always flows in food chains from the first member toward the last member in an upward direction, meaning from producers to consumers all the way to the consumer at the top of the chain. Examples of simple food chains from different habitats are shown as follows:

Aquatic food chain:

▶ Phytoplankton → zooplankton → mackerel → tuna

Woodland food chain:

▶ Grasses → rodent → snake → hawk

Arctic food chain:

▶ Phytoplankton → zooplankton → various fish → seals → polar bear

African savanna food chain:

▶ Grasses → wildebeest → lion

Producers serve as a critical first step in food chains because they capture the Sun's energy. Earth's producers are its photosynthetic plants (grasses and phytoplankton). Primary consumers consist of herbivores: grasshoppers, rabbits, or grazing zebra. Once the primary consumers transfer the plant's energy to animal tissue, secondary and higher consumers continue the flow of energy transfer. Each consumer level above primary consumers contains either a carnivore or an omnivore. Small mammals, small fish, and birds tend to act as secondary consumers, for example, foxes, raccoons, gulls, or freshwater perch and catfish. The top of each food chain contains large predators: white sharks, killer whales, grizzly bears, polar bears, eagles, and owls. Food chains also depend on scavengers such as vultures that feed on carcasses. By doing so, the scavenger exposes the carcass to the air, which hastens decomposition by bacteria and fungi.

The rate at which producers convert the Sun's energy into biomass (any matter of biological origin) is called *gross primary productivity*. Producers first build up enough biomass to meet their own needs for growth, then the extra biomass supplies the food chain. *Net primary productivity*

Ecological Pyramid

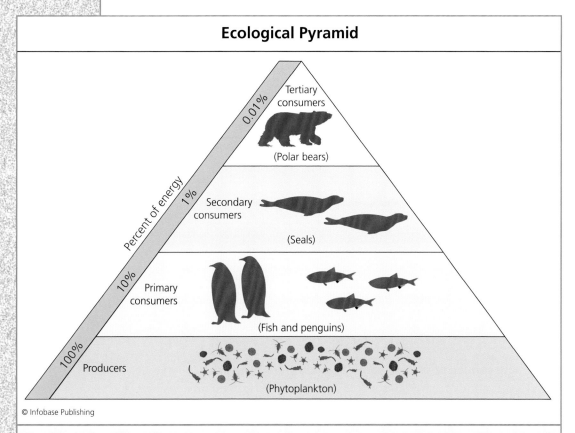

© Infobase Publishing

An ecological pyramid, also called an energy pyramid, illustrates how energy is lost with each step up from energy producers to consumers, and from prey to predators. Top predators adjust to a lowered availability of food and energy by having small numbers of offspring, compared with species lower on the pyramid that produce large numbers of fast-growing offspring.

equals the rate in which producers store sufficient energy to supply food chains, minus the producers' own energy needs, as follows:

gross primary productivity − producer's needs = net primary productivity

Food chains follow the second law of thermodynamics—they progress toward an increased state of entropy, that is, a state of unavailable energy. Secondary consumers recover a fairly large portion of the energy stored in plant tissue, but consumers higher up the chain have access to a smaller proportion of usable energy. The unavailable energy dissipates as heat, and this energy loss puts increased pressure on tertiary and quaternary consumers to meet their own energy requirements. Food chains rarely contain more than four levels because the inefficiencies simply become too great at each step to sustain more levels.

Secondary consumers play two important roles: they convert plant tissue to animal tissue, as discussed above, and they are prey for other species. Prey species usually have large populations for the following reasons: energy is readily available to them, and large population size offsets the amount of individuals caught by predators. Diagrams called *energy pyramids* depict the energy transfers from numerous prey animals in food chains to lesser numbers of predator animals. Energy pyramids also explain why some species are more vulnerable to extinction than others. Even in perfect conditions, predators have much less energy available to them for survival, breeding, raising young, and sustaining a population. These animals, therefore, depend on the well-being of all the biota lower on the chain. As a consequence predators such as eagles, tigers, sharks, wolves, grizzlies, and polar bears become vulnerable to extinction when their environment undergoes damage.

PRESERVATION VERSUS CONSERVATION

People save vulnerable species mainly through *preservation* and conservation. Preservation of natural resources is the act of setting resources aside, unspoiled and undisturbed. Designated wilderness areas, for example, preserve space for wildlife. Conservation is the wise use and management of natural resources for humans' benefit, but in a way that minimizes disturbance to ecosystems. Streams in Montana designated for recreational trout fishing serve the needs of biota and the wishes of people; they conserve nature.

Preservation programs began in this country before naturalists learned of biodiversity. The Wilderness Act of 1964, for example, designates land areas in the United States for preserving the animals living there. Of all types of nature reserves, wilderness areas impose the strictest limitations on human activities and so allow animals to behave naturally and follow their normal breeding patterns.

In 1935 naturalists Aldo Leopold, Benton MacKaye, and Robert Marshall formed an environmental organization called the Wilderness Society and began the first organized effort to set aside vast tracts of pristine land solely as wildlife habitat. Each of these naturalists believed that land must belong to the public rather than be exploited solely for profit, but Marshall put his philosophy to work when he became director of forestry for the U.S. Department of the Interior (1933–37). He once said, "I should hate to spend the greater part of my lifetime in a stuffy office or in a crowded city." From his director's office, Marshall set aside 5 million acres (20,235 km^2) of wilderness, land he as a teen had described as "the woods and solitude." Today the BLM oversees about 400 million acres (1.6 million km^2) of wilderness that permits of little or no human activities. Some human activities do creep in, however, making wilderness preservation one of the biggest challenges faced by protectors of the environment.

Proponents of wilderness development have long argued that the vast amounts of untouched land can sustain logging, mining, oil drilling, and off-road vehicles without affecting wildlife. Since the early 2000s, for example, government officials have targeted Alaska's wilderness for oil drilling. Public policy expert Paul Driessen wrote in a 2005 article supporting new oil exploration in the Arctic National Wildlife Refuge (ANWR): "During eight months of winter, when drilling would take place, virtually no wildlife are present." In addition Driessen argued that drilling in ANWR would actually threaten fewer animals than wind power, which kills thousands of birds each year.

Parts of this country's wilderness now contain activities that Robert Marshall would probably find disheartening. The BLM, for instance, leases land for oil and gas drilling in several million acres of wilderness and the FWS grants right-of-way access for select individuals to enter wilderness areas for business purposes. Industries have used right-of-way permits to build telecommunication lines, pipelines, roads, trails, flumes, and canals. Wilderness also contains casual use, which means industry or scientists may enter undisturbed areas to map new right-of-way routes or collect study data.

Despite the strides of U.S. conservation groups, only about 4.6 percent of land in the United States is protected as part of the National Wilderness Preservation System under the authority of the Wilderness Act. (The IUCN ranks the United States 42nd among all nations in the amount of land protected as wilderness.) Privately run organizations such as the Nature Conservancy, founded in 1951, dedicate labor and money to protect plants, animals, and ecosystems. Indeed private groups may provide a more hopeful future for habitat preservation than government programs because private groups steer clear of business interests better than government leaders. The following table lists the main types of habitat protections in the United States.

The United Kingdom uses a designation called Environmentally Sensitive Areas, which protects agricultural lands that have value because of their wildlife, landscape, or history. These areas may support agricultural activities, but only in a way that protects the land and the biota. Regardless of the type of protection, the following principles should be followed in land conservation:

- large reserves are better than small reserves
- one large reserve is better than several small ones adding up to the same area
- reserves should be close together rather than far apart
- reserve lands should be clumped together rather than strung out in a narrow arrangement
- wildlife corridors should connect reserves
- round reserves work better than other shapes to minimize threats from the edges
- minimally developed buffer zones should surround reserves near populated places

Conservation biology comprises a field of study that helps in designating lands to be protected. Conservation biologists study species and habitat loss, habitat restoration, and biodiversity. Many of their studies rely on aerial mapping with the aid of global positioning systems (GPS) and data-gathering satellites. Satellites map changes within a region over time, and these trends can then be analyzed to determine habitat losses or gains, called *change data*. Change data have been useful in showing

METHODS OF PRESERVATION AND CONSERVATION IN THE UNITED STATES

TYPE OF PROTECTION	DESCRIPTION	MAIN LIMITS ON USE
nature reserve (nature preserve)	protected area for flora, fauna, or nonliving resources	noninvasive research on natural resources
national park or seashore	managed by National Park Service for historical significance, scenic beauty, or wildlife viewing	limits on vehicles; logging, mining, oil and gas drilling, and hunting; fishing and livestock grazing with permits only
wildlife refuge	maintained by government or organization to provide food, water, shelter, and space for one or more wildlife species	active management by government for plant and animal preservation and improvement
wildlife sanctuary	run by private organization to provide permanent habitat, usually created for one or more particular species	allows public to visit for viewing wildlife in natural habitat and behavior
wilderness area	roadless area with very limited human intrusion and no development	hiking, horseback travel, and some scientific study
wild and scenic river	wild, scenic, or recreational rivers protected by the government	wild rivers contain only trails; scenic rivers have some roads; recreational allows swimming, fishing, and boating
national forest	forest land managed and protected by the U.S. Forest Service	vehicles on managed roads and nonvehicle activities in off-road areas

(continues)

METHODS OF PRESERVATION AND CONSERVATION IN THE UNITED STATES *(continued)*

TYPE OF PROTECTION	DESCRIPTION	MAIN LIMITS ON USE
shorebird reserve	landowners conserve crucial sites for shorebird breeding, *migration*, and habitat	wildlife viewing from designated areas
protected wetland	land normally saturated with water for part or all of the year	development activities by the EPA only for restoration and protection from land, air, and water pollution; fishing, hiking, canoeing, and bird-watching only
environmentally sensitive area	managed by organizations to protect flora, fauna, or historical value	no access other than viewing from designated areas
national marine sanctuary	government-managed open ocean and reefs for protecting breeding grounds and natural resources	watercraft that do not disturb marine species with noise or pollution; some scientific study
marine protected area	marine or estuarine area managed by government agencies	removal of species banned or limited
national scenic trail	government- or privately-managed trails of national significance	hiking, horseback travel; overnight use by permit only
research natural area	land *biomes* managed by the U.S. Forest Service, usually inside a national forest	research and education

the effects of logging, roads, and rights-of-way on plant life and animal populations. The World Resources Institute, based in Washington, D.C., operates the Global Forest Watch program, which uses satellite images and geographic information systems (GIS) to assess two components of conservation land: intactness and size threshold. Intactness is the size of a single piece of undisturbed area that sustains natural ecosystems, and size threshold relates to the minimum land area required to support a species' population and allow for its recovery.

Field studies also play a part in conservation biology. Rather than collecting data from high above the Earth, field studies include observations, animal counts, and animal tracking to learn the status of any species' population. Trackers most often use radiotelemetry, a technique in which a single animal in a herd or pack or other social group is caught, tranquilized, and fitted with a collar that emits an electronic signal. After the animal returns to its group, biologists track the group's movements with an antenna that receives the signal. In this way research may be conducted far away from the animals and not disturb the group's behavior.

Conservation biology studies draw on genetics, ecology, natural-resource management, and social sciences. Genetic tools help determine the presence of genes that may give a population of animals a better chance for survival. Ecology lends information on Earth's various regions of vegetation and terrain, called *bioregions,* and how biota and the land work together. Natural-resource management balances the needs of people with the survival of forests, wildlife, and aquatic species. Finally, social sciences, such as anthropology, sociology, philosophy, and economics, play a role in conservation planning that accounts for the needs of humans.

Carrying capacity represents a critical part of all conservation studies. Carrying capacity means the maximum population that a species' habitat can support over a period of time. As population density of any species increases, the capacity of its habitat to support it decreases.

$$\uparrow \text{Population density} \rightarrow \downarrow \text{Carrying capacity}$$

High population density forces more individuals to share a smaller piece of habitat. Food and shelter (nests, dens, burros, etc.) availability decreases, so mortality rates increase, usually for the following reasons: animals conflict with each other more often; increased pathogen and disease transmission; less protection from predators; inadequate shelter for offspring; and stressful interactions, leading to possible rejection from

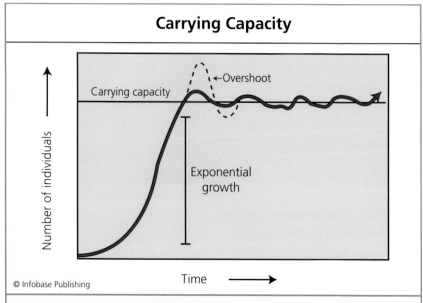

Carrying Capacity

One of the greatest threats to existing natural resources, including plant and animal species, comes from populations that exceed the land's capacity to support them. The human population has grown exponentially from 1800 to today so that it has overshot its resources in many parts of the world. Natural populations of plants and animals adjust to overshoot by producing less offspring, but humanity has not, so today humans use more resources than the planet has the continued capacity to provide.

a protective social group. Species can adjust to dense populations for a time by taking in less food and by producing smaller litters, but these are short-term fixes that weaken the overall population.

When land and food is limitless, animal numbers increase in a logarithmic manner, meaning the numbers increase exponentially. Growth rate then slows as the population reaches the land's carrying capacity. Many terrestrial mammals on Earth may have now reached their habitat's carrying capacity, and they survive because of higher mortality rates and adjusted litter size. When they can no longer make further adjustments, population size plummets, causing what is called a *population crash*. Species living as *opportunists* or *competitors* have a great impact on their population's survival.

Opportunists are species designed to produce large numbers of offspring that receive little or no parental care and so can overcome population crashes. These species reproduce often and the offspring disperse

CARL LINNAEUS—THE FATHER OF TAXONOMY

axonomy is a branch of biology concerned with naming and classifying organisms based on their relationship to each other. The taxonomy in use today originated with Carl Linnaeus (also Carolus or Carl von Linné), born in 1707 in Stenbrohult, Sweden. Young Linnaeus possessed a love of plants, which he put to use as an adult working for Sweden's Royal Science Society. Linnaeus traveled the country collecting specimens for the society and in the process developed a reputation as a skilled botanist. In 1735 he moved to the Netherlands to devote time to the science of plant life and the resemblances he saw between certain seemingly unrelated plants. Linnaeus rearranged plant classifications that had been used in biology since Aristotle and published a new scheme in *Genera Plantarum*. The scheme contained groupings based on detailed likenesses between species rather than gross appearances.

Linnaeus's new hierarchy met with a combination of criticism and indifference from the scientific community, so he retreated to Sweden, married, and became professor of natural sciences at the University of Uppsala. But he continued studying plant life based on similarities and differences in structure, color, reproduction, and other physical traits, and published *Species Plantarum* in 1753. A younger generation of botanists understood the value of classifying plants to the species level, as Linnaeus had proposed. Linnaeus received satisfaction for more than just scientific merit; part of his motivation had come from a desire to honor God's plan by understanding the world of living things.

The hallmark of the Linnaeus system resides in characteristics called *differentia specifica*. These unique characteristics make every organism distinct from all others and enable biologists to classify each new organism they discover without confusing it with others already described. The system additionally made order of all species by grouping them into genera, genera into orders, orders into classes, and classes into two kingdoms. Most significantly, each species can be distinguished by a name that contains both genus and species. Therefore biologists speak of tigers as *Panthera tigris*, an iguana as *Dipsosaurus dosalis*, a redwood tree as *Sequoia sempervirens*, and so on.

(continues)

(continued)

Following Linnaeus's death in 1784, taxonomists paid closer attention to species ancestry. In 1857 Carl von Nägeli proposed that fungi and bacteria be placed in the plant kingdom, but Ernst Haeckel suggested in 1866 that bacteria, protozoa, algae, and fungi belonged to a new kingdom called *Protista* because they seemed related to neither plants nor animals.

Biologists today use nucleic acid sequencing to trace ancestral roots and electron microscopy to study minute differences in cell structures. These methods have enabled taxonomists to alter schemes that had not changed in 100 years. Studies on deoxyribonucleic acid (DNA) and ribonucleic acid (RNA) reveal that all living things belong to one of three domains: Bacteria, Archaea, or Eukarya. Molecular biology may uncover additional ways to classify organisms so that the present system may again change in coming years. Linnaeus's dedication to finding a standard system for classifying biota built a standard for generations of scientists to follow.

quickly. Rats are an example of an opportunist species because they have these traits. Competitor species reproduce late in life, produce few offspring, have long life spans, and one or both parents actively care for their young. Newborn of competitor species tend to be big and are born strong relative to opportunists; their size helps them survive to maturity. Wild horses are an example of a competitor species.

CLIMATE CHANGE AND BIODIVERSITY

Climate change affects the Earth's vegetation by altering normal cooling and warming cycles. On a short-term basis, warming of the atmosphere due to increased carbon dioxide and other greenhouse gases enhances ecosystems. Higher than normal carbon dioxide levels increase the photosynthetic activity, and this makes plants more efficient in using water, perhaps forestalling drought. Over long periods, however, global climate change hurts biodiversity by the mechanisms suggested by the United Nations Environment Programme's World Conservation Monitoring Center and summarized in the following table.

CLIMATE CHANGE EFFECTS ON ECOSYSTEMS	
ECOSYSTEM	**GENERAL EFFECTS**
wetlands	drying out, imbalanced water cycle
coastal marshes	habitat loss in estuaries and deltas due to sea level rise and flooding
tropical forests	drought, invasive species
boreal forests	increased incidence of fires and pests
the Arctic	expansion of boreal forests into arctic habitat, loss of tundra, thawing of permafrost
alpine mountains	habitat shift into higher altitudes, loss of high-altitude habitats, rapid snow melt
low-lying mountains	loss of land to rising sea levels, loss of seabird nesting colonies, human encroachment on habitats
arid and semiarid areas	deserts become hotter and drier, increased desertification, salinization, loss of grasslands and arable land
coral reefs	coral bleaching and death, altered growth rates
mangrove swamps	decrease in area

Climate change alters ecosystems by changing vegetation growth and weather patterns. First, climate change alters the growing season for plants and trees and so affects the species whose breeding cycles depend on these plants and trees. For example, a tree that blossoms at a different time on the calendar affects the insects that carry pollen, and the altered insect populations influence the feeding opportunities for birds and reptiles. Second, new weather patterns that increase the violence and frequency of storms or the severity of heat waves and drought also affect the health of species and their resistance to disease and pests.

Al Gore wrote in his landmark 2006 book, *An Inconvenient Truth,* "... we are facing what biologists are beginning to describe as a mass extinction crisis, with a rate of extinction now 1,000 times higher than the normal background rate. Many of the factors contributing to this wave of extinction are also contributing to the climate crisis. The two are connected. For example, the destruction of the Amazon rain forest drives many species to extinction and simultaneously adds more carbon dioxide to the atmosphere." Many animals already struggle to keep pace with climate change. Polar bears go hungry waiting for the Arctic ice to form from which the bears hunt for seal; fish and crustaceans cannot breed in water too warm to supply their food; and the broods of migrating birds starve because the seedlings and insects they usually eat have come and gone because of changing weather patterns.

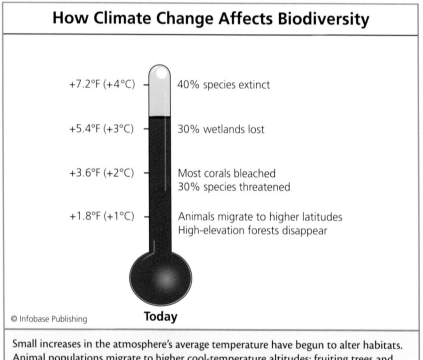

How Climate Change Affects Biodiversity

+7.2°F (+4°C) — 40% species extinct

+5.4°F (+3°C) — 30% wetlands lost

+3.6°F (+2°C) — Most corals bleached
30% species threatened

+1.8°F (+1°C) — Animals migrate to higher latitudes
High-elevation forests disappear

Today

© Infobase Publishing

Small increases in the atmosphere's average temperature have begun to alter habitats. Animal populations migrate to higher cool-temperature altitudes; fruiting trees and plants blossom ahead of schedule to meet the feeding needs of migrating birds; and northern, cold-climate forests become threatened in warmer temperatures. This diagram illustrates some of the changes that have been associated with increased average temperature.

THE GLOBAL COMMUNITY OF SPECIES

Animals do not follow national boundaries when they move about, so agreements between countries become critical to wildlife conservation. The 1975 Convention on International Trade in Endangered Species of Wild Flora and Fauna (CITES), for instance, ensures that commercial sales of wild animals and plants take place in a way that does not hurt their survival. To date CITES protects some 5,000 species of animals and 28,000 species of plants. Unfortunately, the wildlife trade continues to decimate populations in countries that have not signed the treaty. A second major international attempt at wildlife preservation is described in the following sidebar "The Biodiversity Treaty."

Currently 11 countries share 20 of the world's largest preservation sites: Algeria, Australia, Brazil, Chad, China, Ecuador, Greenland, Italy, Saudi Arabia, United States, and Venezuela. South Africa, Kenya, and Namibia provide some of their preserved land for *ecotourism*, a segment of the travel industry that gives people an opportunity to see wildlife in their natural habitats. Ecotourism serves as an educational tool to raise awareness of endangered animals, and it puts money into local economies. Sadly, poachers and the illegal animal trade care little for ecotourism dollars and continue to make their income at the expense of wildlife.

ENVIRONMENTAL ETHICS

Environmental ethics focuses on a difficult question: Which is more important to planet Earth, providing for humans or protecting species other than humans? Some ethicists have tried to answer this question by putting a value on all the world's species, which, of course, is difficult, by dividing the question into three different value systems:

- commodity value—Can an animal or plant be made into a drug, a product, or be sold?
- amenity value—Do species improve human life in a non-material way, as Henry David Thoreau described? That is, Thoreau believed life improved simply by having the opportunity to observe other animals.
- moral value—Do species have value on their own and not related to any human needs?

The world contains thousands of species that humans know little about, and these creatures and plants seemingly have no value at all in the environment. More likely, science has not yet deciphered the roles these species play. How is one to put a value on the thousands of unknown species? Most important, is the value of an entire endangered species worth more or less than a person who is hungry, poor, or jobless?

The Endangered Species Act's well-meaning attempt to halt the disappearance of animals and plants opened a Pandora's box of ethical questions. This law and biodiversity in general are complex subjects, yet humans may be running short on time to solve many of the puzzles held within ecosystems.

THE BIODIVERSITY TREATY

In 1992 the United Nations convened the Earth Summit in Rio de Janeiro, Brazil, and representatives of more than 150 nations came together to discuss the preservation of species. Each nation signed the Convention on Biological Diversity, also known as the Biodiversity Treaty. This treaty called for worldwide listing of endangered and threatened species as well as cooperation among nations for their preservation, including a pay-for-use plan in which industrialized nations pay developing nations for any plants and animals they take.

Despite ambitious goals, the program fell victim to red tape and bureaucracy. Part of the problem arose from a fear that large corporations might exploit developing nations that contain most of the world's biodiversity. Precautions against this exploitation hampered scientists' attempt to study biodiversity in developing nations. Curator Douglas Daly of the New York Botanical Garden told the *New York Times* in 2002, "Something that was well intentioned and needed has been taken to an illogical extreme." Daly and other scientists complained that it had become easier to cut down a forest than to study it.

The Biodiversity Treaty's critics cited three central concerns. First, the pay-for-use plan created an opportunity for rich nations to exploit poorer nations, an activity called biopiracy. Second, developing countries could possibly hold their resources for ransom, and, third, the resources might disappear even faster with the treaty in place. The experience of Professor Ricardo Callejas of the University of Antioquia in Colombia highlights the difficulty of navigat-

CONCLUSION

Biodiversity is the abundance and variety of species in the world or in an area. This simple concept has given rise to laws to preserve biodiversity as scientists began to learn that ever-increasing numbers of plants and animals have been disappearing since humans began dominating the Earth with machines. The Endangered Species Act has some success in protecting species, but it has also been mired in debate and controversy, especially when saving an endangered species conflicts with economic needs. People have a hard time comparing the value of a plant or an animal life to the value of human needs.

As the public learns more from scientists about the way all parts of nature interact with one another, they may understand the difficulty of

ing the treaty's rules. He explained to the *New York Times* the challenges of collecting a small sample of plants for his research: "If you request a permit you have to provide coordinates for all sites to be visited and have to have the approval from all the communities that live in those areas. Otherwise, go back to your home and watch on Discovery Channel the new exciting program on dinosaurs from Argentina. I am still waiting after fourteen months for a permit for collecting [black pepper species] in Choco." Ironically, President George H. W. Bush refused to sign the treaty, claiming it was too vague. Opponents added another worry: The treaty would impede the work of pharmaceutical companies that sought new compounds from unique biota. In 1993 President Bill Clinton acknowledged those worries but agreed to sign the treaty, saying, "We cannot walk away from challenges like those presented by the biodiversity treaty. We must step up to them." As Clinton suggested, the Biodiversity Treaty continues to play a role in conservation.

Today the treaty's signers continue to study biodiversity-related issues such as *global warming*, habitat destruction, poverty, and the business aspects of environmental science. Said Callejas, "I have trouble convincing my closest friends that what I do is because of passion, curiosity, a desire to know more about a group of organisms." Perhaps Callejas's friends feel as many people do who think biodiversity loss has become so enormous a problem, a single treaty cannot fix it.

putting a value on biodiversity. Yet it seems that the only way to save biodiversity is to put some sort of value on it so that individuals and corporations, governments and special interests have an incentive to save it. A very difficult ethical question about the value of biodiversity therefore lies hidden among all the lists of species and habitat maps.

So how do individuals hope to preserve biodiversity? The task requires cooperation between governments and scientists and a good deal of honest communication so that opponents of species protections understand that human life will follow right behind all other life if a mass extinction occurs. Biodiversity studies combine a variety of biological sciences with social sciences, ethics, and philosophy. Perhaps biodiversity will be saved only by understanding the big picture of life's place in the universe. Scientists can then use that understanding as a driving force for their detailed studies on individuals, species, and habitat.

MEASURING SPECIES
AND EXTINCTION

Scientists studying biodiversity must determine the number of species on the Earth and the numbers being lost to extinction. Biodiversity studies therefore rely on solid estimates of the population size of species, which is affected by species behavior, migrations, breeding cycles, and the size and distribution of habitats. All of these assessments can be rather difficult to make, especially when studying biota that live largely mysterious lifestyles in remote places.

Biodiversity measurements involve five main challenges. The first challenge comes from the difficulty of determining the number of species on Earth. Second, molecular biology is constantly revealing more about the genetic makeup of species—how they are related and why they are similar or dissimilar. Sometimes a species does not succumb to extinction, but rather it acquires genes that help it adapt to changes in the environment. Third, the characteristics that certain species require in their habitat may not be fully understood, and this makes it difficult to assess the status of a species when a habitat is altered. Fourth, not all species on Earth have been discovered, and science has no way of knowing whether or when these species go extinct or their role while they inhabit the Earth. Finally, animals often move around. Biologists must determine whether an animal's population is declining or whether the species has simply migrated to a new place. A useful adage in biodiversity studies is the following: "Every animal is rare somewhere."

This chapter describes the different types of biodiversity and the main techniques used in biodiversity studies. It discusses why biologists focus on certain species called *indicators* to gain information on a larger number of species in the same ecosystem. The chapter also covers advances in preservation technology and reviews ethical issues that all ecologists confront as they strive to save species, discussed in the sidebar "Environmental Ethics" on page 41.

TYPES OF BIOLOGICAL DIVERSITY

Biodiversity may be defined in five different ways. *Genetic diversity, species diversity,* and *ecosystem diversity* describe biological systems made of plants or animals, or both. *Ecological diversity* and *functional diversity* describe the Earth's biota in large systems, such as biomes. Most people with some idea of biodiversity tend to think of species diversity, which is the number and variety of species on Earth. Biodiversity actually ranges from the molecular level within a species' genome to a much larger scale that encompasses the entire environment.

Genetic diversity occurs at the molecular level within deoxyribonucleic acid (DNA) and in an individual's genes. Genetic diversity causes differences between individuals within a population and, in many cases, a specific set of genes possessed by one animal enables it to survive when others in its group cannot. The idea of genetic diversity becomes clear by considering specific examples to show how this type of biodiversity gives certain individuals an advantage over others within a population.

- Cheetahs that run faster than others have a better chance at catching prey.
- Bull walruses that are bigger and stronger than other males have a better chance of breeding with females.
- The brightest colored male cardinal has the best chance of attracting females.
- Wildebeests that maneuver the best over the African savanna have a good chance of escaping lions.
- Salamanders colored in the most deceptive camouflage have an increased chance of going unnoticed by predators.

An animal's traits come from its genotype, which is the collection of genes that make an animal look and behave as it does. Gene analysis provides information about adaptations such as speed, strength, and camouflage that enable certain individuals to thrive while others succumb. Genotypes also describe the relatedness of members in a population and between populations. Relatedness may in turn indicate declining population size, interrupted migration routes, destruction of breeding grounds, or disrupted habitat. This is because when populations become fragmented or decline in size, the diversity of the group's members begins to decline; the individuals making up the group become more related to each other over a few generations.

Genetic diversity studies begin by taking tissue, blood, or hair samples from animals. The biologist then determines the nucleic acid sequence of the sample's DNA to define the animal's genotype. Any unique pattern in DNA's gene sequence may indicate a specific trait that gives an animal favorable characteristics. Genotype can be more complicated than this simple description, however. No single gene gives a cheetah its extraordinary speed, but it may be possible to find a finite set of genes that determine a salamander's camouflage pattern. Conservation biologists may one day be able to inject into an animal genes that will confer advantages for the animal's survival. The pathologist John H. Wolfe at Pennsylvania's School of Veterinary Medicine has said in a campus newsletter, "Through gene therapy, we replace a 'broken' gene . . . with the correct, functioning copy." This technology may take a long time to enter conservation biology, but the fast rate of species loss demands that new technologies move quickly to save the world's critically endangered species. Such gene manipulation will also prompt arguments from those who denounce this practice because it goes against nature and the new *transgenic* animal may carry potential and unknown dangers. The debate about genetically modified species continues to unfold, and no conclusions have yet come from these discussions.

Species diversity consists of two important components: richness and *evenness*. Species richness comprises the number of species living in a region or a community; species evenness relates to the abundance of individuals within a select species. Biologists estimate species diversity by manually counting animals in the wild within a specific region, and then illustrating the results on a map of the region. Species diversity maps almost always provide mere estimates rather than exact numbers because most animal species can be difficult to count. One technique to estimate animal numbers involves counting small groups, such as a herd or a flock, and then extrapolating those results to estimate the total population size. Though biology now has accumulated fairly accurate estimates for easy-to-count animals like elephants, rhinoceroses, pandas, or golden eagles, counting secretive species or ones that live in hard-to-reach habitats—snow leopards, whales, white sharks, or birds in jungle canopies—still present unique challenges. Some of the techniques used today for counting animals are covered in chapter 7.

Ecosystem diversity encompasses all of the various ecosystems known to biology. Some examples of different ecosystems are marine kelp forests, coral reefs, the African savanna, and pine forests. Ecosystem diversity also

The forest canopy is a unique habitat that acts as a transition between the terrestrial Earth and the atmosphere. Scientists study the nearly inaccessible forest canopy by using walkways built in the canopy, such as this one at the ExplorNapo Lodge along the Napo River in Peru. Forest canopies, particularly in the Amazon forest, contain extraordinary biodiversity and are home to some animal species that never set foot on the ground throughout their lifetimes. *(ExplorNapo Lodge)*

refers to the variations within ecosystems. For example, a freshwater lake is an ecosystem, but it also contains subecosystems at the lake bottom, at mid-depth, at the water's surface, and along its banks.

Ecology is the study of ecosystems and how they relate to each other. Ecology contributes to the study of biodiversity by helping biologists gain an understanding of biomes, including the types of plants living in them, their terrain, their physical features, and their climate. The health of biomes influences the status of many of the individual species native to those biomes. For example, land that 50 years ago supported grassland contained grassland ecosystems, soil ecosystems, and possibly aquatic ecosystems in ponds and streams. After a prolonged drought due to climate change, for instance, the grassland may turn into a dry plain (desertification) and the grassland ecosystems yield to desert ecosystems. All of these events have a profound effect on the biodiversity in these places and contribute to ecological diversity.

Ecological diversity comprises the variety of forests, wetlands, grasslands, lakes, oceans, *riparian* areas, and other *biological communities* that interact with their living and nonliving environment. (A biological community is the collection of animal and plant populations living and interacting in a particular area. The populations of species that lived on the vast plains of the western United States before the 1800s are an example of a biological community.) Air and ground surveys help biologists assess ecological diversity, which in turn gives a clearer picture of ecosystem diversity.

The final type of biodiversity is functional diversity, which is the variety of biological and chemical processes (called *biogeochemical cycles*) that make energy and nutrients available for biota. In other words, functional diversity provides various ways for the energy-matter cycle to operate.

THE NUMBER OF SPECIES ON EARTH

In the species approach to conservation, biologists focus on saving individual species from extinction by studying and assessing endangered species populations and their habitats. The ecosystem approach includes studies of larger systems so that sufficient areas may be preserved to sustain healthy habitats in general.

The species approach to halting biodiversity loss depends on estimates of the number of species on Earth. The exact number will probably never

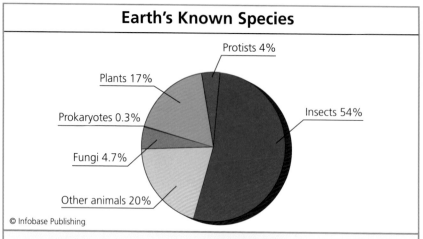

Earth's Known Species

Protists 4%

Plants 17%

Prokaryotes 0.3%

Insects 54%

Fungi 4.7%

Other animals 20%

© Infobase Publishing

Environmental scientists do not know the exact number of species on Earth. Some species have not yet been discovered, and others disappear before anyone discovers them. Scientists have identified more than 1.4 million species, which may represent only 10 percent or less of all species.

be known, but the estimates used today in ecology come from counting the known species, then deriving the total number of unknown species.

The Convention on Biological Diversity has reported that 1.75 million species have been identified so far, and most scientists surmise that the actual number of all species—known and unknown—is at least 14 million. Rather than understanding the number of species on Earth, it may be more important for students of biodiversity to recognize certain hallmarks of biodiversity, as follows:

1. Species diversity increases nearer the equator.
2. Tropical rain forests cover about 7 percent of the globe's land but hold more than 50 percent of species.
3. The loss of all species has accelerated since the year 1800.

Edward O. Wilson, curator at the Museum of Comparative Zoology at Harvard University, is one of society's preeminent biodiversity scholars. In his classic 1988 text, *Biodiversity,* Wilson wrote, "No precise estimate can be made of the numbers of species being extinguished in the rain forests or in other major habitats, for the simple reason that we do not know the numbers of species originally present . . . extinction rates are usually estimated indirectly from principles of biogeography . . . the number of species of a particular group of organisms in island systems increases approximately as the fourth root of the land area. This has been found to hold true not just on real islands but also on habitat islands, such as lakes in a 'sea' of land, alpine meadows or mountaintops surrounded by evergreen forests, and even in clumps of trees in the midst of a grassland." Wilson's thoughts perhaps best sum up the reasons why counting the Earth's species may be impossible.

KEYSTONE SPECIES

Keystone species, by their natural behavior, determine the condition of other species in an ecosystem. If keystone species decline or go extinct, an entire ecosystem may not survive. Wilson has suggested with other ecologists that past extinctions were very likely due to the decline of certain keystone species that, in turn, caused ecosystem imbalance. Ecosystem imbalance occurs when one or more components disappear, putting stress on the remaining members of the ecosystem.

Sea otters and bees are but two examples of the many keystone species found in ecosystems. Sea otters' main food is sea urchins. By holding

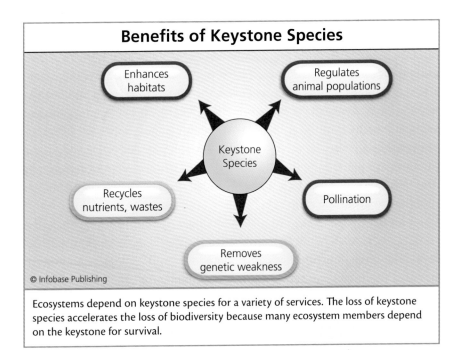

Benefits of Keystone Species

Enhances habitats

Regulates animal populations

Keystone Species

Recycles nutrients, wastes

Pollination

Removes genetic weakness

© Infobase Publishing

Ecosystems depend on keystone species for a variety of services. The loss of keystone species accelerates the loss of biodiversity because many ecosystem members depend on the keystone for survival.

sea urchin populations in check, otters keep them from devouring kelp forests and seaweeds that are home to hundreds of types of fish, perhaps thousands of species overall. Bees benefit ecosystems by pollinating plants and trees, which provide protective cover for other plants and habitat for insects. The insects then serve as a food source for many bird species, upon which foxes feed, upon which raptors feed.

Foundation species play a slightly different role from keystones. They change habitat in a way that enhances it for other species. African elephants play the role of foundation species through their feeding activities in which they crack and uproot trees, which helps open up areas of land and aerates soil. This condition promotes the growth of grasses for grazing animals, and eventually small bushes and woodlands follow the grasslands. The new vegetation favors tree grazers, insects, birds, and the predators that hunt them. Elephant feeding behavior therefore benefits not only ecosystems but an entire community.

In predation, one organism hunts, kills, and eats another organism(s) for its nutrition. Predators are carnivores or omnivores, and their prey are usually omnivores and herbivores. Predators play a keystone role by regulating populations of prey animals and thereby maintaining balance within food webs. Examples of predator keystone species are wolves, lions, alligators, white sharks, and owls.

Predators furthermore keep populations of prey in check in a non-random manner. In other words, they do not usually attack the first prey animal they find, but rather select their prey. Predators remove old, weak, and sick animals and, as a consequence, the remaining strong and healthy population passes on its genes to offspring. Healthy prey populations then support healthier and perhaps larger predator populations. For example, years marked by very large North American deer populations have resulted in above average mountain lion populations. The converse situation may also occur in which as a prey population declines, predators regulate their numbers to avoid starvation.

Another set of relationships in biological communities comes not necessarily from feeding habits, but from other aspects of animal lifestyle. These relationships are called symbiosis, and biological communities contain three main types: *mutualism, commensalism,* and *parasitism.* In mutualism, two or more species benefit from each others' activities. For example, the bacteria living in the gut of termites receive a safe, nutrient-rich habitat, while they provide a benefit to the termite by digest-

Predators play a crucial role in every ecosystem by helping to regulate prey populations, which helps control overgrazing of plant species. The four predators shown here serve in these roles in their distinct ecosystems: (a) red-tailed hawk in New Mexico *(Lin Evans)* (b) snapping turtle being caught for tagging and scientific study *(Mark Corner and Diverse Mobile Outdoor Environment, University of Massachusetts)* (c) African lion *(Herb Coleman)* (d) white shark *(Alice Christie)*.

ing woody food. In commensalism, one species benefits and the other is neither helped not harmed. Redwood sorrel, a small herb that lives in the shade of redwood trees, has a commensal relationship with the redwoods. Shade from the tall trees gives the sorrel its preferred habitat but the redwood tree receives no known benefit. Finally, parasitism is a relationship in which one organism benefits and the other is harmed. As an example, *Phytophthora* fungus lives on oak and other trees as a parasite and the fungus eventually kills the tree.

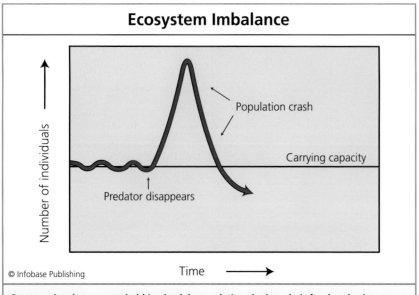

Ecosystem Imbalance

Number of individuals

Population crash

Carrying capacity

Predator disappears

© Infobase Publishing

Time

Prey species that are not held in check by predation deplete their food and other resources, and this leads to stress, competition, disease, and starvation, which are all characteristics of ecosystem imbalance. These circumstances cause a prey population to crash, which then threatens the predator.

The following sidebar "Environmental Ethics" examines the relationships that exist when humans become part of an ecosystem.

GENE POOLS AND NICHES

A *gene pool* is the collection of all the genes in a particular population of individuals. Genetic diversity enhances biodiversity overall by improving the traits carried in a species' gene pool. Therefore today's biodiversity programs rely on the knowledge gained from genetic studies.

Genes control the traits of every plant and animal, and each generation transfers its genes to the next through asexual or sexual reproduction. Sexual reproduction gives an advantage to organisms because it creates more diversity in their gene pool by combining the traits from two unrelated parents. A greater variety of potential parents and greater variety of pairings in a breeding season therefore increase genetic diversity in offspring. Over time an animal population acquires advantages for survival in two ways: variations in its gene pool and random gene mutations.

ENVIRONMENTAL ETHICS

Earliest human society consisted of hunter-gatherers in which members of a settlement ventured afield to collect plants and fruits, to fish, and to hunt meat-producing animals. In these societies, humans behaved as predators in a sustainable manner, meaning they hunted to sustain their village but they did not decimate wildlife populations.

Today Earth approaches 7 billion people, which is beyond its carrying capacity. Population densities in Africa, Asia, and South America have forced some people into a far more menacing predator role in which wildlife numbers and habitat disappear in the face of human activities. Environmental ethicists have confronted the underlying cause of this problem: poverty. The International Union for Conservation of Nature (IUCN) and the European Commission published this viewpoint in an undated briefing titled *Biodiversity in Development: The Links between Biodiversity and Poverty*: "Poor people themselves are often the cause of biodiversity degradation and loss, especially if lack of income alternatives drives them to over-exploit the resources." This statement emphasizes the complex association between human poverty and biodiversity.

Environmental ethics involves the search for a solution for two opposing needs: human hunger and wildlife survival. Many African communities depend on their native wildlife for food (called bushmeat) and for income. On a small scale this practice was at one time sustainable, but an increased demand for food and space far exceeds the capacity of wildlife populations to keep up. Threats to the survival of African wildlife now include the following: people hunting for food, destruction of habitat for agriculture or urban development, disappearance of prey animals, retaliatory or preventive killings to protect villages from predator animals, and illegal wildlife smuggling and *poaching* for income. Some residents sell their native animals as sources for medical drugs, non-medical cures and supplements, aphrodisiacs, religious and ornamental items, and exotic foods. Wildlife protections exist in many nations but, unfortunately, as governments strengthen the protections, black-market prices for animal products soar. As a result, the extinction of some animals hastens rather than slows.

(continues)

(continued)

These forms of legal and illegal hunting have been driven by hunger. Starvation is an immediate crisis for people in many regions of the world in addition to Africa, and things like bushmeat often provide a family with its only protein source. So a need to protect endangered wildlife faces another equally critical need: preventing human starvation.

Some corners of the world rich in biodiversity have revised their relationship with native endangered wildlife. For instance, hunters in Thailand and Costa Rica refrain from capturing rare birds, reptiles, amphibians, or fish because they can make more money keeping them alive (instrumental value) for ecotourism. These two countries now earn more from ecotourism than they would by destroying animal habitats.

Ethicists must also consider conflicts between biodiversity and cultural needs that may not have an obvious instrumental value. The black rhinoceros's habitat in sub-Saharan Africa has stayed about the same size in the past 30 years, yet 90 percent of the animals have disappeared. Two related factors have contributed to this tragedy: a black market that deals in rhino horns and petroleum. In Yemen, young men earn daggers with elegantly carved handles made from rhino horn as a symbol of status and wealth. As oil-rich Yemen's wealth has grown in the past few decades, the country has become the world's largest importer of black rhino horn to make these ceremonial pieces. The Convention on International Trade in Endangered Species of Wild Flora and Fauna (CITES) has put a global ban on the import of black rhino horn, but as a result the black market for rhino horn has flourished. Armed guards now protect many rhinoceros herds against poaching, and conservationists have even resorted to tranquilizing the animals and cutting off their horns to dissuade poachers. The dehorned animals confront another problem: adults use their horns to establish herd hierarchies and to protect their young. As with most ethical dilemmas, more than one aspect exists to each problem.

In 2005 conservationist Adam Oswell said in a radio interview in Australia, "In countries where people don't make a lot of money, they're not concerned about killing animals; they just want to feed their family and make money." The link between poverty and biodiversity cannot be explained any better.

Even a minute change in an individual's genes might give an animal a better chance of adapting to environmental change. Within a few generations, the advantageous gene has been passed on to many of the group's offspring. For example, peppered moth populations living in London, England, changed their coloration during the 1950s from light gray to sooty black. The reason? Each generation of moths had gained genes for dark color so that the moths could blend into a landscape marred by pollution, smoke, and dark soot. This process increased the moths' *fitness,* a species' ability to sustain health and reproduce in its environment. Small genetic changes that enhance fitness in any specific population—London moths compared with the same species in Liverpool—over a few generations is called *microevolution.*

Animals evolve to meet changes in their environment by acquiring adaptations, and these adaptations may also determine a species' role in an ecosystem. This role of a species within an ecosystem is called an *ecological niche,* or simply *niche.* Students often assume that to "occupy a niche" means an animal occupies a specific location. This is actually the definition for habitat; a niche is a species' lifestyle or role in that habitat.

To occupy a niche, a species depends on certain foods, plants, and physical and chemical conditions within its ecosystem. *Fundamental niche* refers to the combination of potential physical, chemical, and biological factors that animal species use for survival. The concept of potential is important in this definition, especially when many niches overlap. Earth's species have evolved to occupy fundamental niches that eliminate competitions, but when a species occupies a niche that overlaps with another species' niche, it has two choices for survival: compete or adapt.

Competing directly with another species may increase the number of deaths in both species' populations, so for the benefit of both, species often adapt to avoid competition. When different animal species adapt in this way, they are said to occupy a *realized niche,* a specialized portion of the fundamental niche. For example, elks in North America digest woody plants that grow on high mountain slopes, as well as plants that grow on flat lowlands. Elks are therefore capable of occupying a fundamental niche as a general grazer. Wolves have evolved to hunt elk on flat terrain to make the best use of the pack's ability to chase a herd and separate out a single individual. Elk therefore increase their survival chances by spending as much time as they can (other than to find water) on mountain slopes, where wolves tend not to hunt. They therefore occupy a realized niche, that of high-slope grazer of woody plants. Wolves benefit too because they conserve the energy it would take to chase elk over mountainous terrain. The wolves can carry out more successful hunts by targeting elk herds that descend for water.

North American elk make a behavioral change to occupy a realized niche, but other animals undergo a physical adaptation to accomplish the same thing. For example, two lizard species may look and act similarly and both may prefer to feed on the same types of insects. In the same habitat these species would compete directly for the same food, but through microevolution one species becomes slightly larger than the other. The larger variety of lizard ingests larger insects and leaves the tiny meals for the smaller lizard. Rather than compete, each lizard conserves its energy by feeding differently instead of competing. This behavior is called *niche differentiation,* or *niche splitting,* and it occurs only between two similar species.

How do niches affect biodiversity? The more specialized the niche, the more vulnerable an animal is to change in its habitat. Conversely, the best survivors, called *generalists,* thrive in broad niches; that is, they survive on many different types of food and tolerate a wide range of environmental conditions: crows, coyotes, cockroaches, and humans live as generalists. The "Case Study: The March of the Argentine Ants" gives an example of what happens when a generalist invades a habitat. Specialists, by contrast, are not as versatile and occupy narrow niches. Specialists tend to live in only one type of habitat, on a single type of food, or in a narrow range of environmental conditions. The northern spotted owl, discussed earlier, occupies a narrow niche. Other examples of specialists are giant pandas, polar bears, tiger salamanders, and red-cockaded woodpeckers. The woodpecker illustrates the precarious lifestyle of some specialists. This bird nests by carving holes only in longleaf pines that are at least 70 years old. Old longleaf pines have become limited to the southeastern coastal plain of the United States from the Carolinas to Louisiana. If the remaining longleaf pine forests disappear, the woodpecker will disappear, too. Meanwhile birds of the family Corvidae, more commonly known as crows, are generalists with no such restrictions. Crows range to just about every landmass on Earth and eat—with only slight exaggeration—anything in sight!

Two distant relatives that may have very different futures. (a) A three-toed sloth hangs in the foliage in a Costa Rican rain forest. This species has a fairly restricted habitat in only certain tree species that live in northern and central South America. *(Keith Sirois)* (b) A raccoon is a generalist that adapts to varied environments throughout North and Central America, parts of Europe, and Japan. *(David Menke, U.S. Fish and Wildlife Service)*

ENVIRONMENTAL INDICATORS

Environmental scientists would learn very little about habitat and diversity by monitoring crows or other generalists. Specialists, however, often fluctuate in numbers in response to changes in the environment. For this reason many specialists are called indicator species because they react dramatically to changes in the environment and so serve as early warnings of environmental decay. The following table lists some indicator species and the information they provide in environmental science.

Like a canary carried into a mine to detect deadly gases, birds serve as harbingers of danger in the environment before humans sense it. Certain species require very specific habitats, so monitoring them is the best way to monitor that habitat, whether it is a wetland, beach, riparian area, woodland, or forest. Birds therefore serve as excellent *biodiversity indicators* for the following reasons:

EXAMPLES OF ANIMAL INDICATOR SPECIES

INDICATOR SPECIES	ECOSYSTEM OR CONDITIONS MONITORED	INDICATOR'S REACTION TO CHANGING CONDITIONS
lichens	air pollution	decreases with increasing sulfur dioxide in air
nudibranch (sea slug)	coastal ecosystems	decreases with loss of prey due to pollution or ocean warming
krill	marine ecosystems	enlarged ozone hole causes decreases, which halts marine food chains
amphibians	land-water habitats	decreases in drought, wetlands destruction, and waterways pollution
frogs	land-water habitats	enlarged ozone hole allows ultraviolet light to kill eggs; tadpoles killed by water pollution; reproductive failure in adults from soil and water pollution
trout	freshwater with high levels of dissolved oxygen	decreases in polluted, *eutrophication*
butterflies	plant habitat fragmentation	decreases with increased fragmentation
birds	habitat loss	decreases in migrating species dependent on habitats for stopovers; decreases in specialist species with habitat loss
black-footed ferret	North American grasslands	disappearance of its food source, prairie dogs, with development of grasslands

- live in every climate and biome
- participate in almost every terrestrial and aquatic ecosystem on the earth
- migrate between climates and biomes
- respond quickly to changes in habitat
- easy to track and count
- give behavioral clues to threats
- most species play a central role in numerous food webs
- different species depend on certain terrestrial plants, aquatic grasses, trees, seeds, insects, rodents and other mammals, and marine species as food
- reproduction is sensitive to pollution

MEASURING SPECIES LOSS

Animal species decline may be estimated using mathematical models, which are built on vast data collections on the number of species found in each habitat. Three complementary methods supply those data: field surveys, global mapping of plant diversity to assess animal habitats, and estimates of animal numbers based on the effect of humans on habitat loss, in a technique called the *biodiversity intactness index* (BII).

Field studies consist of manual counts of animals in the wild followed by the creation of species lists. Local wildlife officials in national parks and university research teams usually do the counting and they compile the data from the study areas to build a picture of species distribution in a habitat or across larger regions such as counties and states.

Species counts provide valuable information on areas containing a high degree of biodiversity, as well as areas with little biodiversity. Areas having the greatest biodiversity are called *hotspots,* and these places have drawn worldwide interest as the most critical for protection. Species counts also help biologists determine the effects of habitats damaged by reduction in size, fragmentation, pollution, and other forms of destruction.

Two methods provide biologists with useful information in addition to that received by manual counts. The first method, camera surveillance, enables scientists to observe animals that live secretive lifestyles or in remote places. The Wildlife Conservation Society currently has set up 15

CASE STUDY: THE MARCH OF THE ARGENTINE ANTS

$\mathcal{S}$ ome species by their mere presence indicate that an ecosystem is under threat. The spread of Argentine ants from South America to North America illustrates how one species directly and indirectly can endanger other species by taking advantage of a quirk in genetic diversity.

Argentine ants arrived in New Orleans in the late 1700s aboard coffee-carrying ships from South America. By 1820 the insects had spread throughout New Orleans, and within the next century the ants began an inexorable march northward beyond Louisiana. By 2000 they had spread into the following states: Louisiana, Mississippi, Alabama, Georgia, Florida, South Carolina, North Carolina, Tennessee, Arkansas, Texas, and Oklahoma. Argentine ants have now spread into the Pacific Islands, Europe, and southern Africa.

Like many successful invaders, Argentine ants display aggressive tendencies when they take over new territory. They fight and defeat native insects and soon overrun natural ant colonies, the hives of native stinging insects, and even birds' nests. Genetic studies of these far-flung populations have turned up a surprising phenomenon that helps them as invaders: The ants have almost no genetic diversity. A single small colony, perhaps from a single ship in the port of New Orleans, may have provided the ancestors for all the generations that have now spread around the globe.

The Argentine ants passed through what is called a genetic bottleneck in which very little genetic diversity exists. Usually such bottlenecks confer a disadvantage on species, but the Argentine ants manage to use their relatedness as a distinct advantage. Normal ant colonies have organized social structures so that members work in a cooperative fashion, and when one colony invades the territory of another, the two families fight for the spoils. Argentine ants, however, act as one big family of close cousins. Because they all recognize each other as kin, they form huge cooperative colonies that overtake species normally able to repel attacks. When Argentine ants invade, they actually make biodiversity go backward because they eliminate all other various ant

cameras in northeastern Cambodia along trails used by rare Asian tigers. As an animal passes between paired cameras, it breaks a detection beam and triggers the cameras. (Two cameras are used for each tiger because tiger stripes are not symmetrical and the combined images from each flank help researchers identify individual animals.) Despite this high-tech approach to species monitoring, cameras have caught only one tiger, but the project leader Ed Pollard dreams of the time when the tiger's habitat is sufficiently preserved to sustain a population. "In twenty years," he said in species and replace the territory with their one supercolony. The biologist Andrew Suarez noted in 2000 in the *New York Times*, "Some people say, 'Big deal; you're just replacing some ants with others.' But the thing is, you're displacing twenty species. And all the roles they play are wiped out: dispersing seeds, pollinating plants, providing food for other animals. All the functions are totally lost when Argentine ants move in." These ants wreak all this havoc by creating what is called a *trophic cascade.*

A trophic cascade occurs when a change in one species' population leads to major disturbances of other unrelated species. California's coast horned lizard and the similar Texas horned lizard both live a life evading their predators: birds, mammals, and snakes. As their defenses, the lizards rely on spiny armor and cryptic coloring to blend in with their desert habitat. Each lizard species also reproduces slowly, so they are not likely to develop adaptations fast enough to meet changes in environment. Enter the Argentine ants. Both lizards depend on native harvester ants as their main food source, but when the Argentine variety decimates native ant colonies, the lizards cannot adapt to new food sources. When the lizards roam out of their natural habitat in search of harvesters, their camouflage does not match the surroundings and predators pounce. Meanwhile plants that depend on harvester ants to spread their seeds also decline. When the native vegetation declines, so too do insects that depend on the plant life. The birds that eat the insects disappear, and next the animals that prey on birds go hungry.

The biologist Ted J. Case of the University of California at San Diego has offered some words of comfort on the unusual ability of Argentine ants to form huge, homogeneous colonies. The ants' relatedness, he explained to the *New York Times*, "may be useful in the short term, as it is for Argentine ants and some other invasive ants, but in the long run it appears to be an evolutionary dead end because social behavior could not evolve." Meanwhile the Argentine ants demonstrate the intricate ways in which biodiversity can be threatened.

2007 to the Associated Press, "people will come here and drive along the road and there's a distinct chance there'll be a tiger trotting along in front of you." This of course represents the goal of every conservation project.

The second method, fossil records, provides fewer technological results than remote-controlled cameras but offers a historical view of species rise and decline. Fossils give evidence of ancient organisms as well as a physical description of these organisms. Fossil records do not, however, give scientists all the information they need today about biodiversity because today's available fossil records may represent only about 1 percent of the species that have ever lived on Earth.

Global mapping techniques are used for building databases on plant diversity, which in turn provide clues to the likely animal diversity in specific habitats. Any confined study area may be mapped by combining field study data with high-resolution satellite images. On a larger scale, global plant diversity mapping depends on statistics to predict where biodiversity is rich and where it is scant.

Despite the varied manual and technological means of assessing biodiversity, putting a quantitative value on biodiversity remains a difficult task, and no single method does the entire job. For this reason the Convention on Biological Diversity introduced in 2005 a tool for describing biodiversity in a single, standard manner. This BII quantifies the abundance of a diverse population relative to a well-studied reference species. The BII requires only a few weeks to gather data, compared with previous databases on species that have taken decades to compile. Biologists then calculate an index value as a percentage and using three pieces of information: the current way an ecosystem is being used (wilderness, agriculture, ranching, etc.), the area of the land being measured, and the species richness in that area. Using the grasslands of Kansas as a hypothetical example, a BII of 80 percent means that, when averaged against all plant and animal species in the region, Kansas's grassland populations have declined to 80 percent of their preindustrial (about the year 1800) numbers.

The BII assesses general trends in biodiversity rather than exact values, so that biologists learn about *functional groups* like insect-eating birds, small burrowing mammals, large grazing mammals, or rain forest amphibians. The BII also takes into account the types of human activities on the land, which of course greatly impact the status of the species living in the region. In summary, the BII indicates where biodiversity is disappearing the fastest and where it might be increasing.

The BII may soon supplement the information already gained from field studies, fossil records, and mapping. To date, these methods have together shown the following trends in biodiversity:

- Populations have declined 84 percent since the preindustrial period.
- Mammals declined the most in this period, 71 percent.

- Species associated with a specific habitat declined the most, 26 percent, in grasslands.
- Ninety percent of habitat loss leads to the extinction of about 50 percent of the habitat's species.
- Greatest biodiversity losses in Africa are in countries with the continent's greatest population density: Lesotho and Swaziland.

Georgina Mace, director of science at the Institute of Zoology in the United Kingdom, explained to *Nature* magazine in 2005, "Biodiversity assessments need to move away from species lists and species extinction rates, because often the existence and proximity of local [human] populations matters more. [The] biodiversity intactness index makes a start in satisfying many requirements and provides a robust, sensitive and meaningful indicator." For this reason comprehensive indicators such as the BII give the best picture of biodiversity losses and gains.

ASSESSING HABITATS

Preserving species in the wild requires preservation of the species' habitat as well. Abundance or scarcity of habitat plus the species' characteristics determines how species distribute across the Earth, and this distribution follows six different patterns. The first is continuous distribution, in which an animal exists almost all over the world. Crows, which are generalists, are an example of continuous distribution. Second, a disjunctive distribution takes place when a species must live in a highly fragmented habitat. The Indian tiger once lived in a continuous band extending from northern China, along the eastern portions of Asia, and over the Indian subcontinent. Today the tiger's disjunctive habitat includes small, isolated pockets in Southeast Asia. The third type is restricted distribution, in which fragmentation has not occurred but the total area of the habitat is greatly reduced. The mountain gorilla has been restricted to two national parks in Africa, the last habitat available to this endangered species. Fourth, an evolutionary distribution describes animals that were once dominant in their habitat but have now lost most of their population. The scattered survivors of the American bison are an example of this type of distribution. Fifth, a climate-affected distribution is one in which species have been affected by climate change. In some areas of the world, their habitats have been severely damaged or completely destroyed, such as marine coral habitat. The "Watching a Species Disappear" sidebar on page 58 describes another effect of climate on a specialized habitat. Finally, the sixth pattern is endemic distribution. Endemic distribution, called *endemism,* defines species found only in one place on Earth, but they may be abundant and not threatened in that place. Wallaby species found only in Australia provide an example of endemism.

Certain species, called *flagship species,* serve as a symbol of a habitat and are fairly easy to monitor. These species help biologists gauge habitat health and the potential risks to other species living in the habitat, and they act as public relations ambassadors. Conservationists rely on flagship species to symbolize the plight of all animals in the environment, because the public recognizes these animals and may feel more inclined to help them avoid extinction. The Mother Goose Syndrome describes this type of thinking in which people feel a closer relationship to animals they perceive as friendly or cuddly (pandas, otters) and less connection with animals that are dangerous or frightening (snakes, vultures).

Flagship species are usually large animals that people identify easily and represent certain traits in people's minds. Therefore lions are rulers, panthers are stealthy, pandas are cute, and so forth. Conservation biologists also refer to these species as charismatic species, perhaps best represented by polar bears. Global warming and rapidly melting polar ice have created a flagship story for the environment in the past decade. Polar bears need stable ice floes for hunting seals, and, therefore, for their survival. Polar bear images adorn greeting cards, stuffed toys, and many other items in Western society, so people worry more about these animals' fate than they might worry about the other species dependent on the ice. The Canadian ecologist Ian Stirling described to the Associated Press in 2008 a study he led on the effect of global warming on polar life other than bears: "What we wanted to do was look at the whole picture because there's been a lot of attention on polar bears. We're talking about a whole ecosystem. We're talking about several different species that use ice extensively and are very vulnerable." Indeed, an ecosystem contains many roles or niches. The following table summarizes the main types of species, in addition to flagship species, used in habitat assessment.

Habitat assessments often uncover one of the main threats to species today: habitat fragmentation. Fragmentation results from incursion of humans into animal habitats, causing the habitats to break into smaller,

ANIMAL SPECIES USED FOR ASSESSING HABITATS		
TYPE OF SPECIES	**ROLE IN HABITATS**	**EXAMPLES**
flagship	their activities help other species living in the same habitat	tiger, rhinoceros, great apes, marine turtles, giant panda
keystone	in small numbers yet play role in which many other species depend for survival	alligator, prairie dog, wolf, bees, white shark
foundation	help shape and maintain the physical habitat	elephants, beavers, woodpeckers, earthworms
indicator	serve as early warning that a habitat or ecosystem is being degraded	trout, butterflies, frogs, migrating ducks

isolated areas. Some of the major human activities that fragment habitats are roads, sound walls and border walls, fences, canals, pipelines, cultivated land, and suburban areas. Industries, too, such as strip mining and logging fragment habitats, and natural events such as floods, earthquakes, erosion, and lava floes can also cause fragmentation. Any large, continuous habitat may suffer from these occurrences, but the habitats now under the greatest threat from fragmentation are forests, riparian areas, ocean dunes, and open grasslands.

Fragmentation causes two threats to the species living in them. The first threat comes from a separation of a population into two (or more) smaller populations. If animals cannot reach one another, they cannot congregate and breed, so the species loses genetic diversity. Over time mutations and natural selection in two separated groups of the same species make them so dissimilar they can no longer successfully breed with each other. Two geographically separated species then arise from a common one, an event called *speciation*.

Deforestation by clear-cutting destroys large areas of forest habitat within days. In the Amazon forest shown here, people clear-cut forests to increase area needed for cultivation or grazing cattle or for obtaining wood for heating. Clear-cutting may also occur when corporations set up mining, logging, or large-scale agriculture. *(Rhett A. Butler/Mongabay.com)*

The second problem caused by fragmentation comes from the *edge effect*. Edges in the environment are places where two habitat types meet—a forest meeting an open meadow, for instance. Some species thrive on edges because the edge increases their chances of getting food. But edges may also harm species in the following ways: increased threats from predators that hunt along edges, vulnerability to parasites not found deeper within the habitat, threats from harsh weather conditions—wind and temperatures tend to be more extreme at edges—may threaten newborns and decrease the density of insects for insect-eating species. Edges also open up opportunities for some species. Brown-headed cowbirds prefer living along edges where they can act as brood parasites, meaning they infiltrate other birds' breeding areas. The female brown-headed cowbird invades a nest when a parent is away, lays her eggs next to the native eggs, and then flies off. Cowbird young have particularly large, gaping mouths, so native parents bringing food back to the nest tend to feed them more than their own offspring. In the worst circumstances the native chicks starve to death and the cowbird chicks thrive. Fragmentation therefore causes a change in ecosystem behavior.

EXTINCTION

Animals and plants go extinct for many different reasons, but human activities in the last two centuries have had the greatest impact. Since the industrial revolution, the human population has grown exponentially. As a consequence, habitats have been destroyed, fragmented, or polluted, while ecosystems have been injured by climate change, pollution, and invasive species. Poaching, illegal hunting, and trade in exotic wildlife have contributed to the downward spiral of vulnerable species.

To be sure, other factors contribute to extinction, and many of these are natural events that would happen even if humans did not populate the Earth. Fossil records show that some species followed a path of evolution to a point in which new life forms emerged and transformed habitats. The original species did not adapt to the new conditions and so vanished. Species may also go extinct due to natural disasters, natural climate change cycles, competition, and overspecialization. Any extinction caused by things other than a natural event is called a *premature extinction*, because it occurs at a faster rate than would naturally take place. Premature extinctions occur mainly because of the following five human-related factors:

- habitat loss, destruction, or fragmentation
- invasive species
- human population growth and expansion
- pollution
- overharvesting (overhunting, overfishing, poaching)

The above factors plus natural events cause three different types of extinction: local, ecological, and biological. A local extinction may occur when a species no longer inhabits a certain area, even though it can be found in other parts of the world. Ecological extinction means that so few members of a species remain, the species can no longer play its normal role in ecosystems. Finally, biological extinction occurs when a species no longer exists on Earth.

Animal extinctions that have taken place in human history have mirrored the migration of humans from Africa and Asia to Europe, and then to the Americas. Some of the species known to have disappeared during humans' westward migration are the following: the dodo, great auk, passenger pigeon, dusky seaside sparrow, Carolina parakeet, Steller's sea cow, and aepyornis (elephant bird).

Scientists estimate 99.9 percent of all species that have ever existed are now extinct. Some animals succumbed to mass extinctions, which occur periodically over a course of millions of years. Other extinctions take place more frequently, about every few to several centuries, and are called *extinction spasms,* such as the disappearance of dinosaurs. Most biodiversity experts agree on extinction rates of between 0.1 percent and 1 percent per year throughout the world. To put these numbers in perspective, at 0.1 percent, Earth loses 14,000 species per year if there are a total of 14 million species; at 1 percent, Earth loses 140,000 species of the 14 million. Furthermore, at a rate of 1 percent lost per year, at least one-fifth of the world's plants and animals could be gone by 2030 and half by 2100. Perhaps most troubling of all, an extinction rate of 0.1 to 1 percent may be a conservative estimate, which means the extinction rate could be significantly faster. Edward O. Wilson has pointed out perhaps the main obstacle to finding a true extinction rate: "The vast majority of species are not monitored at all." Scientists clearly cannot study things that they do not know exist.

Fossil records help fill in the gaps regarding species that have disappeared. Fossils consist of mineralized sections of things that once lived: bones, teeth, entire skeletons, shells, leaves, or seeds. Sometimes the biological matter has degraded and only an impression in sediment remains, but these impression fossils are valuable because they show what an animal or plant looked like millions of years ago. In addition fossil structures compared with skeletons of present-day animals or plants shed light on how a species may have evolved.

Fossil studies have drawbacks that make them useful only as complements to other technologies. First, some species may have left no fossil records, or, second, their fossils have not yet been discovered. Third, many fossils decompose to a condition that yields little information about the creature that had lived millions of years ago. Despite these disadvantages fossil records have been used to show the following animals became extinct in prehistoric eras: mastodon, mammoth, wooly rhinoceros, saber-tooth cat, and dire wolf.

Three additional technologies supplement the information from fossil examinations. The first technique is radioisotope dating. An isotope is a form of an atom that spontaneously emits an alpha or beta particle or gamma rays. American chemist Willard Libby of the University of Chicago won the 1960 Nobel Prize in chemistry for developing radioisotope dating, which measures the isotopes emitted by fossils. Libby devised a method based on the radioisotope carbon-14, which the atmosphere makes continuously when neutrons from the Sun bombard nitrogen molecules. In one reaction, a Sun neutron displaces a neutron from nitrogen and emits carbon-14 and a proton (N is nitrogen, C is carbon, and H is hydrogen):

$$_7N^{14} + {_0}n^1 \rightarrow {_6}C^{14} + {_1}H^1$$

A small portion of Earth's carbon is carbon-14, which, when it decays, returns to $_7N^{14}$ by emitting an electron. Carbon dating makes use of these reactions plus the two following pieces of information: carbon-14 decay has a half-life of 5,730 years, and living organisms contain a constant amount of carbon-14 (about 1 atom of carbon-14 per 10^{12} atoms of normal carbon-12) in their cells. When an organism dies, it no longer metabolizes carbon, and its carbon-14 content gradually decays. Carbon-14 degrades at a known rate of 693 disintegrations/hour/gram of total carbon, so an estimate of the age of an object becomes

$$t = [\ln (N_f/N_o) / (-0.693)] \times t_{1/2}$$

where ln is a natural logarithm, N_f/N_o is the percent of carbon-14 in the sample compared with that in living cells, and $t_{1/2}$ is the half-life of carbon-14. When it is put all together in the following example, a fossil containing 5 percent of its total carbon as carbon-14 can be dated as follows:

$$t = [\ln (0.05) / (-0.693)] \times 5{,}730 \text{ years, or}$$

$$[(-2.996) / (-0.693)] \times 5{,}730 = 24{,}772 \text{ years old}$$

A second method used in dating fossils involves deoxyribonucleic acid (DNA) analysis. Analysts compare the DNA recovered from a fossil with the DNA from a similar present-day organism, and they use the

WATCHING A SPECIES DISAPPEAR

About 22,000 polar bears live on and near the Arctic Sea ice in northern Canada, Greenland, Norway, and Russia. This flagship species reigns atop the frozen habitat's food web, composed of seals, fish and sea birds, krill and small crustaceans, and finally zooplankton. Polar bears are the world's largest terrestrial carnivore (775–1,400 lb. [350–635 kg]), but today their average weight is decreasing due to a lack of food and a melting habitat.

Climate change has raised temperatures in the Arctic about five times faster than warming in the rest of the world. As a consequence, the Arctic Sea ice that builds and thickens each winter has decreased year by year, leaving less platform on which polar bears hunt ringed, harp, and bearded seals. Less ice means longer swims across open water for the bears that are already receiving fewer calories. In 2005 the marine biologist Charles Monnett alerted the world to the problem at the top of the world. He told Britain's *Sunday Times*, "We know short swims of up to fifteen miles are no problem, and we know that one or two may have swum up to 100 miles. But that is the extent of their ability, and if they are trying to make such a long swim and they encounter rough seas they could get into trouble." Meanwhile, hungry bears on land have started to wander inland to find food; some have broken into houses in northern Canadian villages. In 2006 Interior Secretary Dirk Kempthorne proposed listing the polar bear as threatened on the

assumption that mutations occur in DNA over time at a constant rate. By this method, scientists have estimated the age of Siberian permafrost to be 50,000–100,000 years old.

The third method of dating ancient life is done by core sampling of glacial ice. This technique has helped scientists correlate the depth of the core sample to the number of years ago in which the ice formed. The U.S. Geological Survey's (USGS) National Ice Core Laboratory has recovered cores from polar sites that date to at least 100,000 years.

PRESERVATION TECHNOLOGY

Advanced technologies aid in today's conservation programs like those at the San Diego Center for Conservation and Research for Endangered Species (CRES). As the largest zoo-based conservation center in the world, CRES develops technologies for helping endangered animals survive and reproduce in the wild or propagate in captivity. The table on page 60 endangered species list, reasoning, "We are concerned the polar bears' habitat may literally be melting." Computer-generated models used in polar research today predict the sea ice will have completely melted by 2080 due to global temperature increases.

Polar bears have long confronted other challenges from hunting to the accumulation of toxic chemicals in their bodies, yet 25,000 bears still roam the Arctic and may represent a healthy population. In May 2008 Kempthorne nevertheless announced that the polar bear would receive protections under the Endangered Species Act. This represents the first listing of an animal whose numbers have not yet begun to decrease and also the first listing related directly to global warming.

Opponents of the polar bear's new protected status come mainly from oil interests; drilling companies eye the Arctic as a potential new oil source. M. Reed Hopper of the Pacific Legal Foundation voiced just such opposition of the listing: "Never before has a thriving species been listed nor should it be." Kempthorne admitted to the difficulties of listing the bear as protected, and at the same time explained the polar bear's uncertain future: "When the Endangered Species Act was adopted in 1973, I don't think terms like *climate change* were part of our vernacular." No one knows at this time whether the polar bear's habitat will soon include tankers, drilling rigs, and possible oil spills or remain an undisturbed spot at the top of the globe.

describes the technologies used by CRES and similar conservation groups for preserving both animals and plants. These technologies make up five main disciplines: sustainable populations, bioresource banking, wildlife health, habitat conservation, and restoration biology.

Conservation biologists work for the day they can release wildlife into its natural habitat. After conservation technology, biologists turn their attention to monitoring the animals by manual tracking and counting of animal populations. Various techniques have been adapted to hard-to-study habitats and animal lifestyles, and these methods range from basic data collection on clipboards to instrument-aided monitoring. Radiotelemetry is one instrument method that has become widely accepted for monitoring rehabilitated or released animals.

Radiotelemetry played a role in the recovery of California condor populations from the brink of extinction. In 1987 biologists captured AC9, the last wild California condor, and brought it to a specialized breeding center

(continues on page 62)

STUDY AREAS IN CONSERVATION BIOLOGY

STUDY AREA	DESCRIPTION OF THE SCIENCE
Sustainable Populations	
population monitoring	methods: counting total numbers, counting breeding pairs, and taking fecal and hair samples for genetic analysis of each subpopulation
	goal: determines increase or decline of species populations
reproduction research	methods: collecting information on breeding, offspring numbers, offspring survival, embryo transfer, and artificial insemination
	goal: relates reproductive success to habitat's availability of food and nest/den sites and studies ways to improve reproduction
genetic monitoring techniques	method: collecting samples for DNA analysis, sequencing
	goal: determines differences between and within populations for learning about breeding patterns, histories of individuals, and chromosome features linked to poor reproduction
animal communication system	methods: recording vocalizations at ranges within and outside human detection
	goal: determines how pairs recognize mates, how parents communicate with offspring, and how populations interact
Bioresource Banking	
cryopreservation	methods: collecting endangered semen and tissue for ultracold storage
	goal: makes semen available for artificial insemination programs and tissue available for genetic studies
native seed collection	methods: collecting and storing native plant seeds
	goal: provides for rejuvenation of endangered habitats
DNA bar coding	methods: DNA analysis of bushmeat and exotic species
	goal: aids law enforcement against poaching

STUDY AREA	DESCRIPTION OF THE SCIENCE
Wildlife Health	
disease diagnosis technology	methods: PCR and DNA sequencing of fecal samples goal: determines incidence and spread of infectious diseases and parasites
genomic studies of disease	methods: DNA analysis and chromosome studies goal: identifies genes associated with disease risk, congenital diseases, and immunity
new pathogens	methods: PCR and DNA sequencing of fecal and blood samples goal: identifies emerging viral and bacterial diseases and parasites
disease ecology	methods: PCR and DNA sequencing of fecal and blood samples goal: monitors new disease threats to endangered populations
Habitat Conservation	
telemetry and mapping	methods: geographic information systems (GIS) combined with aerial mapping techniques and telemetry data from collared animals goal: tracks distribution of species in threatened and fragmented habitats and specific habitat use
flagship species studies	methods: telemetry, photography, mapping, collaring and banding, and manual counting goal: tracks declines and distribution of flagship species and enables to assess conservation programs
genetic population studies	methods: DNA analysis of large populations and sequencing to determine mutations; DNA hybridization goal: estimates population sizes over time and identifies genetic vulnerability; identifies how closely individuals are related in populations

(continues)

STUDY AREAS IN CONSERVATION BIOLOGY	
(continued)	
STUDY AREA	**DESCRIPTION OF THE SCIENCE**
Restoration Biology	
release programs	methods: transferring of eggs, nestlings, young offspring, and adults to natural habitat and tracking with various monitoring techniques
	goal: repopulation of natural habitat with threatened or endangered species
translocation programs	methods: release of adults into new natural habitats and monitoring
	goal: determines dispersal, settlement patterns, stress response, and survival rates

(continued from page 59)

at the San Diego Zoo's Wild Animal Park. At the time only 27 condors remained in North America and all lived in captivity. Fortunately, these birds bred well in captivity, and by the early 1990s adults had raised and nurtured several healthy chicks. Human handlers trained the youngsters to avoid power lines and taught the adults to avoid picking up trash items, such as bottlecaps, glass, and shell casings, for feeding to their nestlings. (Condors naturally pick up bits of earth to bring back to their young as a calcium source.) Finally, mentor birds taught the youngsters how to evade danger and to interact with flocks in the wild.

Rescuers returned AC9 to its natural habitat on May 1, 2002. Biologists followed AC9 along with other new releases, and in 2004 a pair of released birds successfully fledged the first condor chick in the wild since the capture-release program began. Today's population of California condors totals more than 300 birds, living along California's coast range, the Grand Canyon in Arizona, and Baja California, Mexico. Many continue to send signals from their transmitters to relieved scientists miles away. At the close of 2007, nature writer John Moir wrote for *Birder's World* magazine, "Not long ago, I saw AC9 myself, a magnificent bird soaring over the chaparral-covered mountains of his ancient home, his mighty wings a symbol of hope that the majestic species will survive." The California condor program exemplified the worth of conservation biology for capture, protection, and safe release into the wild of species on the brink of extinction.

CONCLUSION

Scientists today rely on a blend of different techniques to measure biodiversity, biodiversity loss, and extinction. Advances have enabled environmental scientists to study species from the population level all the way down to the level of genes that give species survival traits. Perhaps the most difficult aspect of quantifying biodiversity comes from the fact that no one knows exactly how many species exist on Earth. Many of today's estimates put the total at about 14 million species, but this is a very rough estimate. Scientists also have only theories on the number of species that have emerged, prospered, and gone extinct by natural means or due to humans.

The status of species in the wild most often results from the condition of their habitat. Healthy habitat indicates that food webs function efficiently and all members of the web have sufficient resources for food, shelter, and raising young. Damaged habitat, by contrast, contributes to biodiversity loss. The main causes of habitat damage are size reduction, fragmentation, destruction, and pollution; all of these problems result from human activities.

In the United States, species in danger of extinction appear on the endangered species list managed by the FWS. The United Nations Environment Programme controls another, more extensive listing of all known species on its Red List, which monitors the status of critically endangered species through species that are abundant and under no immediate threat. Both the endangered species list and the Red List contain plants and animals; the Red List also includes species estimated to have already gone extinct.

Biologists use a wide variety of techniques to assess habitats and species populations. These methods range from on-the-ground studies in which teams count individual plants or animals to sophisticated electronics that track animals and take images of habitats. Sound monitoring techniques enable environmental scientists to focus on specific problems related to biodiversity. Some biodiversity threats are obvious, such as the takeover of an ecosystem by an invasive species. Others are obvious but complicated, such as the myriad effects of human activities on ecosystems. All of these methods and the data they produce have one advantage in common: They produce data that scientists can use to alert the world of a coming crisis in biodiversity loss.

PROTECTING NATIVE FROM INVASIVE SPECIES

Invasion of natural habitats by nonnative species contributes to biodiversity loss and extinctions. Native species consist of plants or animals that normally thrive in a particular community. Invasive species, by contrast, do not naturally belong in the environment they invade; they are also called nonnative, alien, or foreign species. *Introduced species* are those that have been deliberately or accidentally released in a new, nonnative environment. *Exotic species* also arrive deliberately or accidentally, but these species tend to be rare, unique, and highly valued. An orchid known to grow only in Brazil would be an exotic species in Connecticut for instance. By any name, these plants and animals often create biological havoc in the community they invade. This invasion of new habitats can occur by three different means: migration, deliberate introduction, and accidental introduction.

Invasive species threaten native species by crowding the native populations out of their space and using up their water and food sources. They do this by outcompeting the native species in the natives' own habitat, perhaps aided by any foreign parasites that the invader also brings into the community. In many cases, the native species have no defenses against an unfamiliar pathogen or parasite, and their numbers plunge due to disease.

Invasion's main threat to biodiversity resides in ecosystem imbalance, in which normal food webs and other interrelationships cannot work because one or more of the ecosystem's components have been eliminated. Native species in invaded habitats disappear for the following reasons: forced migrations to new areas, disease, starvation, and unnatural behaviors. Behavioral changes often include conflicts between native animals and the invader, or against other members of the same species. Invaders usually bring aggressive mannerisms with them into their new territory, and, as a consequence, native species can offer little resistance to the takeover.

This chapter describes the threats to ecosystems from invasive species and gives examples of invasive species causing trouble in North America. It reviews the series of events that occur after invasion begins and how these steps lead to endangerment and premature extinction. It also reviews programs in which scientists deliberately introduce species into a habitat to return balance to an ecosystem. A special case describing the controversies of such a *reintroduction* is discussed in a study of Yellowstone National Park's wolf population.

ECOSYSTEMS INVADED

Nonnative species may be viruses, bacteria, protozoa, algae, fungi, multicellular plants, or multicellular animals. Many familiar plants and animals in North America began as nonnatives that humans introduced centuries ago. For instance, most of today's cultivated crops and domesticated animals were brought onto the continent by early settlers as food and sources of nonfood products, such as wool and leather. Other nonnative plants served as ornamental additions to gardens, and nonnative animals came as companion animals. These imports arrived in North America from Europe, Asia, and South America centuries ago, but on occasion unwanted imports came, too.

For centuries the holds of ships have carried new animals to new lands: insects, rodents, fish, reptiles, amphibians, and aquatic crustaceans. These travelers disembarked at the same time that settlers and explorers set foot on new continents. Similar accidental releases of nonnative organisms take place today when oceangoing cargo ships transport nonnative organisms in their holds. Many of today's most troublesome invaders have been introduced into aquatic ecosystems when ships empty their ballast tanks in each new port at the end of a transcontinental trip. Parasites and pathogens also get into new environments with infected crew members and tourists.

Invasive species threaten all animals in a habitat, but they are of special concern when they threaten endangered species. Usually the nature of the invader is more suited for success than the endangered species

Endangered hawksbill turtle populations are declining in the Atlantic and Pacific Oceans. Adults live in reef habitats where they feed on sponges, clearing space for other reef species to settle. Hatchlings must complete a harrowing migration from their nests on the beach to the water. Gulls and invasive raccoons, rats, and even dogs feed on thousands of hatchlings every year during this time. *(Ron Mesessa)*

because, unlike many endangered species, invasive species adapt readily to new conditions. In other words, they are generalists, while today's endangered species tend to be specialists. Invasive species hold an advantage over endangered species because they often thrive in places affected by a high degree of human activity, and at the same time they outcompete native species for habitat. The major characteristics of invasive species that lead to their success are the following:

- fast growth due to high reproductive rate and short generation time
- long-lived
- easy means of transport or natural dispersal
- no natural predators (predator animals, pathogens, or parasites) in the invaded habitat

- high genetic variability
- ability to adapt

Adapting to new environmental conditions becomes a vital survival tool for any species, not just invasive species. Native species that have a generalist lifestyle withstand invasion better than native specialists. That is, adaptability and the ability to live as a generalist give an animal its best chances for survival. This becomes especially important when the presence of an intruder changes conditions within an ecosystem. When this occurs, native generalists may be able to adapt to new food sources, new breeding grounds, or new shelters for their newborn.

Even generalists sometimes cannot overcome the most aggressive invasive species. Some invasive species simply overwhelm habitats or ecosystems so quickly that native species have no time to adapt. For instance, fast-growing invasive plants can take over a lake so quickly that other plant and animal life simply cannot keep up with changes in nutrient supply, water temperature, light penetration, or any other factor introduced by the invader. As a consequence, the energy-mater cycle falters as normal producer plants disappear, then animals that feed on these plants suffer. Predators soon decline because they have less prey for maintaining their populations. An entire food chain becomes disrupted and in turn upsets the ecosystem's natural food web.

Invasive animals differ from invasive plants in the way they upset food chains. Animals tend to affect the food chain at the point where they enter it. For example, the European wild boar came to Hawaii about 400 C.E. with Polynesian sailors, and explorers brought these hardy, prolific breeders to the North American continent in the 1500s. Now called feral pigs, these animals thrive in the southern United States from the Carolinas to the west coast. As feral pigs prey on native rodents, populations of large raptors and other mammals that eat rodents lose one source of nutrition. At the same time, the pigs devour natural plant life and uproot the earth in foraging, causing soil erosion. Invasive plants, by contrast, almost always enter food chains at the producer level and may or may not affect the organisms below them in the food chain.

In 2008 the Nature Conservancy scoured databases on ecosystems worldwide to determine the extent of invasive species. Their research revealed that every habitat in the world's temperate climates had some degree of invasion, but the world's most invaded habitat is California's San

Francisco Bay. Several ports line this aquatic ecosystem, and each year more than 3,000 ships plus millions of tourists arrive from all over the world. The invasive species probably enter the bay by a number of conveyances such as the following: on the hulls of oceangoing freighter ships, in ballast water (extra water carried in large ships to provide stability, then emptied when the ship enters a port), on the hulls of pleasure craft, attached to recreational equipment such as kayaks and wind-sailing boards, and even when people discard aquarium water into streams that lead to the bay. Local marine biologist Andy Cohen told KQED television in 2007, "Ballast water is currently the largest mechanism bringing organisms in. From our studies, somewhere between half and 90 percent of organisms now arriving in San Francisco Bay seem to be arriving in ballast water." Ballast water probably accounts for most of the spread of aquatic plants and animal life to other parts of the world as well.

More than 250 species now live in the bay, and as many as 90 percent are thought to be invaders. Though not all of San Francisco Bay's invasive species harm its ecosystems, many do, and the overall effect on biodiversity can be devastating. Cohen added, "Before these arrived, the flora and fauna of the bay were different from what we find in bays in other parts of the world. Increasingly now, as we go from one environment to the other, we see the same organisms in one place after another, and so what we have is this homogenization of the world's flora and fauna." The following table of San Francisco Bay's major invasive species gives examples of how invasion harms this and other ecosystems in other bays around the world.

Sometimes environmental factors conspire to help invasive species build a stronghold in their new territory. First, invasive species very often play the role of *pioneer species.* Pioneer species are rugged first-colonizers of an environment, usually situated low on food chains. Bacteria, fungi, algae, mosses, and lichens provide examples of pioneer species. Higher organisms also act as pioneer species when they are the first of their kind to inhabit an ecosystem. By being the first of their kind, pioneers enjoy an advantage, as competitors, predators, or disease or parasite carriers. Some pioneer species bring additional characteristics that help them gain an advantage in a new habitat. First, even when a pioneer species could serve as food for native *predator species,* the native predator may simply not recognize the pioneer as a potential meal and leave the invader alone. Second, most invasive species succeed in climates similar to their native climates, and this helps reproduction and rearing offspring without a need

EXAMPLES OF HARMFUL INVASIVE SPECIES— SAN FRANCISCO BAY–SAN JOAQUIN DELTA

INVASIVE SPECIES	PROBABLE PLACE OF ORIGIN	THREAT TO NATIVE SPECIES
Animals		
Amur River clam	Russia	devours phytoplankton
green crab	Black Sea	eats shellfish, including native oysters
Asian clam	Asia	eats the food needed by baby salmon and striped bass
Chinese mitten crab	Asia	clogs rivulets and drainage pipes
New Zealand nudibranch (sea slug)	New Zealand	carnivorous predator
Black Sea jellyfish	Black Sea	eats small crustaceans, larvae, and fish eggs
Japanese goby (yellowfin goby)	Japan, Korea, China	outcompetes native fish
hydrozoans	Black Sea	devour small crustaceans, copepods, and crab larvae
Plants		
Scotch broom	Europe	squeezes out native shrubs
Atlantic cordgrass	East Coast	dominates pickleweed marshes and native cordgrass; destroys California clapper rail habitat

to acclimate. Third, some invasive plants ward off competition from native plants by depositing chemicals into the soil that inhibit the growth of the native plants.

SPECIES COMPETITION

Species competing for living space, water, and food is called *interspecific competition*. When resources are abundant in a habitat, competition is minor or does not take place at all, but when resources become scarce, interspecific competition becomes a critical factor in determining whether an endangered animal lives or dies. Invasion may cause resources to become so scarce, in fact, that members of the same species begin to compete to stay alive, an event known as *intraspecific competition*. In addition to food, water, and space, plant or animals also compete over the following things in both healthy habitats and invaded habitats: sunlight, grazing land, soil nutrients, migration corridors, nesting sites, den sites, shelter, and hiding places.

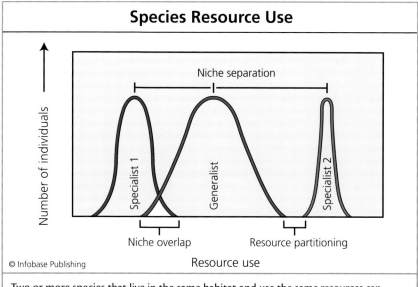

Species Resource Use

Number of individuals

Niche separation

Specialist 1

Generalist

Specialist 2

Niche overlap

Resource partitioning

Resource use

© Infobase Publishing

Two or more species that live in the same habitat and use the same resources can coexist without competing. Specialists do not have much flexibility in resource use, but generalists can adjust their uses to avoid competition. Niche overlap occurs when two species use the same resource in identical ways. No overlap results in resource partitioning, which is advantageous for the strict Specialist 2 in this diagram.

Severe direct competition between species causes extinction of one of them in a situation known as *competitive exclusion*. In competitive exclusion, one species' population outcompetes another species' population vying for the same space and food, leading to a local extinction of the loser. Perhaps the two species simply cannot devise a way (by adapting) in which both can thrive in the same ecosystem. Usually the more adaptive species wins, and the loser is then considered the less fit organism for the ecosystem. The defeated species must then migrate to a new habitat or become extinct.

Migration may not favor some animals, however, because it leaves them vulnerable to new environmental stresses, diseases, and predators. For that reason some animals have devised ways to avoid competition altogether. This avoidance serves two purposes: They save energy by not competing, energy that can be used for other daily activities, and they avoid mass migrations. Two means by which animals avoid competition are by sharing resources or by finding a unique niche.

Resource partitioning allows two different animal species to share the same resource in a slightly different manner. For example, wolves and coyotes compete, yet they find ways to share the habitat by focusing on different hunting techniques. Coyotes often hunt alone and for small mammals; wolves hunt in packs to prey on much larger animals such as elk. Though each species can subsist on the other's main food source, they focus on the resource less likely to interest their competitor. The features of this tactic are discussed further in the "How Resource Partitioning Preserves Species" sidebar on page 76.

Other species avoid competition by seeking a unique niche in a process called *niche specialization*. In niche specialization, two species develop specialties that allow them to occupy separate niches in the same ecosystem. As they evolve toward these distinct niches, they reach a point in which neither species need compete directly with the other. For example, plovers and sandpipers are birds that both live along beaches. Plovers have evolved large eyes and short, strong beaks for finding crustaceans on the sand's surface. Meanwhile, sandpipers hunt the same sands as the plovers. Sandpipers have much smaller eyes than plovers so depend less on sight for hunting, but they do possess much longer beaks that probe deep into the sand for food the plovers cannot reach. The following table describes the discrete adjustments that species make to share an ecosystem or a habitat.

SPECIES SURVIVAL METHODS		
METHOD	**CIRCUMSTANCE**	**EXAMPLE**
adaptation	to adjust to changes in habitat due to external factors or competition	pepper moths adapted to sooty smog (London, 1950s) by turning from light gray to dark gray in successive generations to hide from predators
competitive exclusion	one species outcompetes another for food, water, and habitat	European starlings outcompete other cavity-nesting birds for nests, and so drive out (exclusion) the native birds
migration	a species avoids inhospitable conditions or competition by moving to a different habitat	locusts conduct mass one-way migrations in Asia and Africa when population peaks and food becomes scarce
niche specialization	two species avoid competition by sharing a niche in specialized ways	Plovers and sandpipers both prey on beach crustaceans; plovers feed in shallow sand, while sandpipers feed in the same sand, but deeper
realized niche	a species occupies a limited niche even though it can occupy a larger, more general niche, usually to avoid competition or predators	elk are capable of grazing on mountains or on flatlands, but they occupy a realized niche as mountain grazers to avoid predators
resource partitioning	species with similar needs divide up the resources in an ecosystem, usually when resources are scarce	hawks and owls both eat rodents, but hawks hunt during the day and owls hunt at night

The associations between different species represent symbiosis, which is a cooperative relationship between dissimilar species; it may be thought of as the opposite of competition. Symbiotic relationships may be between two animals or they may be plant-animal, plant-plant, microbe-animal, or microbe-plant. Two main types of symbiosis between species in ecosystems are mutualism and commensalism. In mutualism, two species interact in a way that benefits both. Bacteria living inside an animal's intestines provide an example of a mutualistic relationship because the bacteria receive nourishment while helping the host digest food. In commensalism, one species benefits and the other is neither helped nor harmed. When barnacles attach to gray whales, they catch food as the whale swims, but the whale is unaffected. Parasitism occurs when one species benefits but the other is harmed, such as ticks living on a coyote.

MIGRATIONS

Animals migrate to escape floods and fires, human activities, competition, or invasion from other animals, or when habitat reaches its carrying capacity. Migration is thus a natural way for animals to avoid starvation and find land that provides resources to sustain their population. Competitive exclusion causes some migrations when one species, often a generalist, outcompetes another species, often a specialist. If one species excludes the other from the habitat forever, this is called *one-way migration*. In the animal world, one-way migration is sometimes the only way to survive when an aggressive nonnative species invades a habitat.

Migrations aid the environment in four ways. First, grazing herbivores contribute to the health of grasses by grazing and by fertilizing the soil during migration. Second, when migrating birds or fish leave an area, insects, flowering plants, plankton, small fish, and invertebrates receive time to reestablish their numbers. Third, migrating animals often restore balance to ecosystems they enter by controlling prey numbers or providing food to predators. Fourth, migration contributes to genetic diversity within a species by allowing members of groups to mingle, split off from the herd, or join the herd.

Migration patterns consist of the times, distances, and routes taken, called *migration corridors*. Migration corridors comprise strips of land, air, or water where a migrating group travels unimpeded, and they are influenced by three things: resources for food and shelter, habitat destruc-

Caribou migrations are among the largest migrations in the Western Hemisphere. These caribou in Alaska's Togiak National Wildlife Refuge migrate in herds, behavior characteristic of animals that must eat while they watch for predators during their migration. Habitat fragmentation and blocked migration corridors present very dangerous threats to migrating species. *(Aaron Collins, U.S. Fish and Wildlife Service)*

tion along the route, and population density—both animal and human. Migrating animals need places along their routes to rest and eat before continuing their trek. (Some species migrate nonstop, however, over remarkably long distances.) Migration corridors provide space for the migration's numbers, safety from predators, and food.

Many migration corridors have been destroyed or blocked, and biologists have noted a decline in migratory species in the ocean, in air, and on land. Pipelines, roads, housing developments, and walls and fences all act as barriers to migrations. (Roads are a critically hazardous obstacle to California's endangered tiger salamander.) When the United States–Mexico border wall was proposed, Christine Haas of the Arizona Audubon Society

HOW RESOURCE PARTITIONING PRESERVES SPECIES

Nature uses resource partitioning as a clever way to allow species seeking the same resources to thrive side by side in the same habitat. Rather than trying to build a relationship in which they split the habitat's resources in equal halves—humans have a difficult time doing this also—each species evolves traits for using the resources in a way that will not threaten its competitor. For example, hawks and owls both live as predators at the top of their food chain and hunt rodents (rats, mice, voles, squirrels, etc.), small mammals (skunks, rabbits, gophers, etc.), small birds, snakes, and amphibians. To partition these food resources, hawks hunt during the day and owls hunt at night.

Resource partitioning comes about not by choice but through a genetic change in the species so that it develops a trait favorable in its new niche. This occurs over a time period long enough to allow natural selection for favorable traits, a process called adaptation. Adaptations may be physical features, physiological processes, or behavioral characteristics. In 1859 British naturalist Charles Darwin described the process by which species either survive through natural selection or disappear because they cannot adapt to changes in their environment. In his book *On the Origin of Species*, Darwin wrote, "Though Nature grants long periods of time for the work of natural selection, she does not grant an indefinite period; for as all organic beings are striving to seize on each place in the economy of nature, if any one species does not become modified and improved in a corresponding degree with its competitors, it will be exterminated." Competition initiates natural selection, which gives rise to adaptations that in turn enable resource partitioning.

Individuals that develop traits allowing them to partition resources occupy a new niche for their species, and the next generations may do the same. Eventually an entirely new species, or subspecies, fills the new niche. This process of filling new niches over generations is called *adaptive radiation*, and it is the main reason for biodiversity on Earth today.

said, "If the wall goes up, it will be a complete and absolute barrier for terrestrial wildlife." Haas's prediction has played out at the border fencing in the Sonoran Desert that now blocks migrations of jaguar, pronghorn, and owls. Biologist Emil McCain told *Audubon* in 2007, "It's [the border fence] probably the finest example of habitat fragmentation you can think of." Other human-caused actions have exerted subtle effects on migrations, such as nighttime light emitted by cities that upsets bird migrations, which take place at night. This disturbance is called light pollution or photopollution.

The instincts that guide animals along their migrations have evolved over the millennia, yet Princeton University ecology professor David Wilcox stresses that migration in these times represents an extraordinary "act of faith." Migrating animals do not know what might have become of their wintering grounds, spring nesting sites, or stopovers that have provided food and water for centuries. If migration corridors continue to erode, animals will have one less option for survival.

CHARACTERISTICS OF ADAPTIVE SPECIES

Adaptation occurs when an animal develops one or more new traits through natural selection. These changes in the animal's total genetic makeup, called its *genome,* come about by mutations in deoxyribonucleic acid (DNA). Evolution occurs on Earth today as it did for millions of years because of mutations, but some animal populations respond to environmental changes in an accelerated manner called *microevolution.* Microevolution refers to any set of small genetic alterations in a community's population that allow the population to adjust to changes in the environment. Macroevolution, by contrast, is long-term development of a new species, like the evolution responsible for populating Earth.

Favorable adaptations pass from parents to offspring until the new traits become part of the genetic makeup of all successive generations. In a threatened habitat, the individuals most likely to survive are those possessing the advantageous traits. Animals and plants develop the following adaptations to gain a measure of protection against predators: camouflage, poisons, warning colors, smells, stingers, thorns, quills, bad taste, and mimicry. Microbes also adapt protective mechanisms, but they do this much quicker than multicellular organisms because microbes need only a few hours to produce a new generation; thousands of generations from a single microbe can grow literally overnight. The current problem of bacterial resistance to antibiotics illustrates the adaptive nature of bacteria. Doctors began using antibiotics to treat infections in the 1940s, and within 20 years numerous pathogens had developed resistance to these antibiotics. The majority of today's common pathogens are resistant to almost all antibiotics.

Adaptive species in the animal kingdom possess characteristics enabling them to remain healthy and reproduce in spite of damaged habitats, noise, traffic, pollution, or other disturbances. In many ways, the characteristics of adaptive species resemble those of invasive species: Adaptive species breed prolifically, produce many offspring, grow fast, possess genetic variability, and are generalists. Generalist adaptive species live in many different habitats, tolerate wide ranges in climate, eat a variety of foods, and resist disease and parasites. Familiar examples of adaptive species found living in proximity to people are cockroaches, flies, mice, rats, raccoons, coyotes, and crows.

The American crow is one of the most adaptable generalists in the animal kingdom. Crows are scavenger omnivores, meaning they eat plants and animals and also feed on carrion. Some inventive crows have been known to stand on a sidewalk, waiting for traffic to roll over acorns in the road. The crow then saunters out to pick up an easy meal. (Jason Finley)

Humans are generalists, adaptive, and also invasive. People live in virtually every place on Earth, exist on a variety of diets and in a wide range of living conditions, and can adjust their behavior to avoid predation, pollution, and disease. To withstand changes in the environment, humans simply invent a new technology that ensures their survival. Humans are predators in a number of food chains, and they invade habitats because they proliferate to large numbers rapidly and outcompete native species. Finally, humans play the role of pioneer species when urban areas encroach into pristine wilderness. Mountain lions, grizzly bears, wolves, lions,

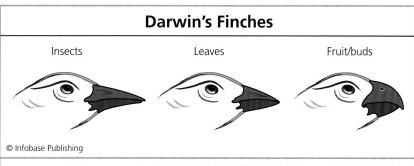

Darwin's Finches

| Insects | Leaves | Fruit/buds |

© Infobase Publishing

The Galápagos Islands were one of many places that Charles Darwin visited to observe the world's wildlife. Several subpopulations of finches have been associated with Darwin's studies because they demonstrate the principles of evolution and adaptation. In general, the size and shape of related finches differ depending on the islands on which they live and the type of food they eat on each of those islands. Of about 13 different types of Galápagos finches, sharp-billed varieties probe for insects in tree bark, medium-billed birds catch insects on leaves, and stout-billed finches crack open cactus seeds as their main diet.

and sharks are all predators that naturally steer clear of humans. Rare attacks from these animals on humans usually involve mistakes or because the animal feels that a person threatens it or its young. The familiar phrase "He is more afraid of you than you are of him" describes the reaction of most natural predators when confronted with the novel sight of a human.

Adaptations having the most effect on biodiversity are changes that lead to adaptive radiation. Charles Darwin noted in the Galápagos Islands in 1835 that characteristics of one bird species was common to all the islands. The birds—ornithologists later identified them as finches—seemed to be of the same species, yet each had slightly different-shaped beaks, depending on the island on which they lived. Darwin had surmised that an ancestor of the finches had once lived on one of the islands, but occasionally visited the other islands. Darwin wrote in his 1839 memoir *The Voyage of the Beagle,* "One might really fancy that, from an original paucity of birds in this archipelago, one species had been taken and modified for different ends." Darwin had drawn his conclusion from his observation that all the islands' finches possessed slight variations in beak size and shape to master the food in their local habitats. "Darwin's finches," as they became known, adapted to their local habitat, yet they shared the entire archipelago without interfering with each other's well-being.

THE CONSEQUENCES OF INVASION

Invasive species interfere with food webs and disrupt ecosystems. The many interdependencies among species in an ecosystem can be radically upset by an invasion. If an invader eliminates one, two, or more of the members of a web, the entire web can potentially disappear.

Biodiversity provides a hedge for Earth's biota from a variety of assaults. Ecosystems have always been buffeted by natural occurrences such as extreme heat and cold, storms, natural disasters, and epidemics, but they withstand such onslaughts because of the diversity contained within them. Ecosystems survive disruptions according to what was described in 1955 by Princeton University biologist Robert MacArthur as the Diversity-Stability Hypothesis. According to the hypothesis, less complex food webs with fewer interactions between species are more vulnerable than more complex food webs with more interactions between species. This is because in simple food webs each predator's livelihood depends on a small variety of prey. If the predator's habitat becomes threatened, it has

The great horned owl is such a skillful predator it might also be considered an invasive species. It dominates almost all of North and Central America and is moving south through South America. Many less adaptive owls disappear from territory invaded by the great horned owl. In addition to other bird species, the great horned owl eats reptiles, amphibians, fish, insects, invertebrates, and mammals such as rodents, rabbits, skunks, and cats. *(Gary M. Stoltz, U.S. Fish and Wildlife Service)*

few alternatives for survival. Complex food webs, by comparison, contain greater diversity of species and more alternatives for food chains to function. In other words, complicated ecosystems withstand threats better than simple ecosystems, which can also be thought of as sensitive ecosystems. Complex ecosystems possess what is called a *buffering effect* against potentially harmful environmental changes. The sidebar "Zebra Mussels" examines a situation in which native species lack a buffering effect and cannot adapt to an invasion.

ERADICATION

Eradicating invasive species from a habitat requires hard work with sprays, poisons, and backbreaking labor plus another obstacle: people's opposition to eradication programs. Few people mourn the loss of a million or so zebra mussels from a lake, but they feel differently about hunts to reduce invading deer, pigs, golden eagles, hawks, rabbits, or foxes.

Eradication programs use chemical, physical, or biological methods. Chemicals have been tried on the Great Lakes zebra mussels, but it is difficult to attain a high enough concentration of any chemical that will not also injure native species. An eradication program in northern California's Lake Davis has been used to remove northern pike released by fishermen more than a decade ago. The pike devour other fish in the Davis watershed along with frogs, crayfish, and ducklings. Biologists have tried the chemical rotenone in a long-term program to rid the lake of the pike, but the poison also kills all the native fish. After eliminating the pike, biologists then must restock the lake with native fish. Residents near Lake Davis have been leery of this type of chemical eradication, especially since the pike have evaded more than one attempt to wipe them out. In 2007 *USA Today* quoted Bill Powers, mayor of a nearby community, as cautioning, "I agree with the scientists who say there's no safe levels of carcinogens in drinking water," when he discussed possible contamination of the aquifers near the lake. Store owner Tammy Milby gave her opinion on the economic harm caused by the pike's invasion of one of California's favorite fishing lakes: "Hopefully it'll work this time. I don't know if the community, the businesses, can handle another failure." Powers and Milby both express the concerns attached to eradication programs.

Physical eradication consists of hunting and trapping. Trapping programs capture invasive animals so biologists can humanely release them into a less sensitive habitat. Sometimes, however, exotic animals require speedy capture because they have entered an environment that does not suit them, and leaving them in the habitat would be inhumane. People

ZEBRA MUSSELS

Zebra mussels (*Dreissena polymorpha*) are small (fewer than 2 in. [5.1 cm] long), striped mussels native to the Caspian Sea. In the late 1980s a transatlantic cargo ship entered the Great Lakes and emptied its mussel-infected ballast water into Lake St. Clair. The mussels stayed there until 1992 but have now moved into all five Great Lakes plus the Huron, Mississippi, Tennessee, Ohio, and Hudson Rivers.

Like other invasive species, zebra mussels are prolific breeders: A single adult female produces between 30,000 and 1,000,000 eggs per year. The tiny larvae live as planktonic creatures, meaning they drift suspended in waters rather than settle to the bottom. This characteristic helps them travel in currents, and once they reach a new body of water, the young or the adults are not choosy about the aquatic ecosystem they invade, usually with human help. Invasive species expert Jim Carlton told *National Geographic* in 2005, "Before humans started moving around, the rate of species movement was a geologic rate. Now we're moving species faster and farther than they ever would or could have moved in nature." Adult mussels attach to surfaces and then secrete a strong, plasticlike fiber that sticks to living and nonliving things. The mussels then get transported from one body of water to another on boats or anything else that picks them up in the water.

Zebra mussel invasion creates five main ecosystem problems. First, mussels efficiently filter water for food and remove almost all the water's algae and phytoplankton. In other words, they act as *exploitative competitors* because their actions remove an entire component from a food chain. Second, with phytoplankton gone, crustaceans, small fish, larger fish, and higher predators in the food chain each lose their main food source. Third, clear algae-free water allows more aquatic plants to grow, which benefits some species such as dabbling ducks but harms other aquatic species by crowding them out and blocking sunlight. Fourth, the mussels concentrate toxic pollutants in their bodies then excrete the pollutants in their feces, endangering native plants and animals near the mussels. Invaders that produce toxins harmful to ecosystems are called *interference competitors*. Fifth, zebra mussels have decimated the Great Lakes population of American unionid clams by attaching to their shells and restricting their movement, breeding, and feeding. The native clams are now near extinction in the Great Lakes, and it is likely that zebra mussels will cause similar harm to other native species.

release reptiles, amphibians, tropical fish, and tropical birds intentionally or accidentally into places that endanger the animal. Owners of exotic animals do not always understand the special needs of these pets, and

Zebra mussels have also made a surprising contribution to the ecology of the upper Midwest. Some migratory duck populations have rebounded because they have adapted to eating zebra mussels to fortify themselves during migration stopovers. Fish such as catfish, sunfish, sturgeon, salmon, and yellow perch have also adapted to eating the mussels. Finally, smallmouth bass numbers have risen because the young have abundant aquatic plants in which to hide and grow.

The zebra mussel invasion has put an economic burden on local businesses because of the cost of removing mussels from boats, ships, buoys, docks, and piers. The mussels have clogged the intake pipes of factories, power plants, water utilities, and other businesses that withdraw water from lakes and rivers. These places must now mussel-proof their pipes by installing strainers that keep out the intruders.

Zebra mussels are marine and freshwater invasive species that provide an example of how globalization contributes to the movement of species around the world. These animals were brought to North America from Eurasia in cargo ship ballast. Scientists have since learned that ship ballast is a major means of transport of invasive species. *(ECHO, Lake Aquarium and Science Center, Leahy Center for Lake Champlain)*

Jim Carlton summed up the issue of invasive species like the zebra mussel, which have migrated halfway around the world in ballast water: "Five thousand or more species could easily be in motion on any given day." Though invasive species move around the planet today with ease, the problem is not new. Carlton explained in a 2005 *National Geographic* article, "Fouling is responsible for quite a few [in San Francisco Bay], especially from the gold rush, when so many old, heavily fouled wooden ships sailed in and were abandoned here. The oyster trade. Fish bait. Ballast water brought the rest." New Zealand, Norway, and parts of the United States have now put legal restrictions on how and where ships may empty their ballast tanks, but that will not affect the zebra mussel invasion that has already spread.

the animals die soon after being released because they are not adapted to the climate, food, or other conditions in an unfamiliar environment. Though some exotic animals can go extinct for these reasons, others cause an opposite problem: They aggressively invade the new habitat. Swamps, woodlands, and grasslands have all been victimized this way by exotic species. Like other invasive species, exotic species prey upon native animals, threaten keystone and flagship species, and consume plants or prey needed by others in the natural ecosystem.

Hunting programs consist of tracking and killing invasive animals. Hunters licensed by the Fish and Wildlife Service (FWS) or private contractors hired by government agencies perform the eradication. Many people oppose hunting on ethical grounds, especially if leg-hold traps or dogs are used to catch the animals. Eradication is also criticized when people have become accustomed to an invasive species and now consider them regular inhabitants. In 2005 a heated debate ensued over the hunting of feral pigs on the Channel Islands off California's coast, brought there by farmers in the 1850s. Animal-protection advocate Scarlet Newton told a local university newspaper, the *Daily Nexus,* "The island pig has no malice. It is just gently looking for food and is not causing a fraction of the damage humans have caused. The best solution is to leave the ecosystem alone. It is as stable as it can be and still be vital." Eradication by hunting will likely remain a very controversial way to stabilize animal populations.

Wildlife experts have also begun to use biological eradication in which they put a species into an imbalanced ecosystem with the intent to combat an invader. Vineyards, for example, contain only one type of vegetation (call a *monoculture*) that upsets the natural ecosystems of the original woods, meadows, or hillsides. Vineyards often draw gophers, ground squirrels, feral pigs, and birds that feed on vines, uproot plants, or build dens. Many vintners encourage natural predators to enter the area as an alternative to chemical poisoning, hunting, or trapping. Sustainable vineyards have relied on foxes, coyotes, and raptors living in the area to deter invaders.

ANIMAL REINTRODUCTION

Natural-resource management uses animal reintroduction for the following two purposes: to restore balance to a damaged ecosystem and to return an endangered species to the wild. In many cases a reintroduced species can turn the tide on invasion, especially when eradication gives only partial success. Rather than put effort into eradicating difficult invaders, biologists can reintroduce animals that can restore balance to the ecosystem.

Reintroduction is done by either relocating animals from another community where their population is dense or by releasing them from breeding-in-captivity programs. Each method requires careful management of the animals during their capture, transport, and release. Reintroduction programs are more than trucking a group of animals to the wild and opening their cages. Many animal advocates have concerns about reintroduction because they believe the process puts too much stress on both the released animals and the native species that have learned to live in their habitat without the added competition. Other critics fear for the safety of people, pets, and domesticated animals. The following case study (on page 87) on the return of gray wolves to Yellowstone National Park sheds light on this controversy.

Wildlife biologists develop a thorough understanding of the habitat before reintroducing any animal. Captivity breeding programs such as that run by many zoos have programs that prepare animals for release and ensure the best results for all members of the natural community. The Association of Zoos and Aquariums provides guidelines on the best ways for carrying out a reintroduction that is successful both for the reintroduced animal and for the habitat's existing species. According to the guidelines, a reintroduction should satisfy the following conditions:

1. aid a native population that can be self-sustaining once it is in the environment

2. preserve the existing ecosystem

3. take place in the animal's normal habitat

4. take place in an area of the animal's normal distribution

5. occur only in habitats not threatened by destruction

6. protect the reintroduced animal from aggression from animals in the habitat

7. not be a solution for disposing of surplus captive animals and orphans

8. not threaten the survival of species already living in the habitat

Mentor animals that live in captivity also help by teaching other members of their species about finding food and avoiding predators. Handlers teach young birds how to avoid power lines, and all activities at the training center try to minimize any connections a wild animal might develop with humans.

Ideal reintroduction habitat should be free of overhunting, land development, fragmentation, pollution, exotic predators, or diseases that could infect the reintroduced species. Large animals such as wolves usually enter in numbers of 50 to 150 to help them build a social structure. Biologists furthermore try to release animals that are similar, genetically and physically, to the natural populations that once lived in the area. Veterinarians meanwhile make sure the animals are healthy and have no injuries; they are vaccinated before their release. Animals with infirmities or behavioral problems do not qualify for release in order to prevent them from being harmed in the wild.

Transport should involve as little noise and other stress as possible. Biologists prepare mammals for release by holding the animals in pens for a period of time at the release site. This acclimation period gives the animals time to overcome any stress caused by the trip, adjust to climate, recognize landmarks, and develop relationships in their group. Biologists provide food for certain released species, such as condors, to help their transition into new surroundings, while other species, such as wolves, simply leave their enclosure to begin their new life in the wild. Many reintroduced animals have transmitters attached to them so scientists can monitor them by radiotelemetry and learn about their behavior.

Even after all these precautions, animal reintroductions can be a tense time for biologists because no one can be certain of the success or failure of animals they have trained and come to know. Geneticist Hans Peter Koelewijn of the Netherlands summed it up this way at a conference on wildlife reintroduction in 2008: "You can have a technical strategy, a scientific strategy, and a socioeconomic strategy, but the animals also have their own strategy." Biology contains very few absolutes.

PLANT REINTRODUCTION

Animal diversity depends on the plants that form the foundation of each food chain and define each of the Earth's biomes. Trees and plants act as producers in the energy-matter cycle, so they are vital to animal life.

(continues on page 90)

CASE STUDY: YELLOWSTONE'S WOLVES

Few animals are as reviled as the wolf. In myth, legend, and literature, wolves have symbolized a deadly force that intends to harm humans. Reintroduction of gray wolves to Yellowstone National Park has had to overcome not only scientific obstacles but society-based hurdles.

North America's first settlers lived among a gray wolf population of 380,000. In the 1800s landowners began developing the west and hunted wolves to prevent attacks on cattle, sheep, goats, horses, and even their families. Some hunters stalked wolves with such dedication for 50 years that by the time Yellowstone Park opened in 1872, the park contained all the animals it has today except for wolves. In the 1930s the federal government sought to finish the job by supporting a wolf-eradication program of poisoning and hunting. Within 10 years the animals' territory had dwindled to a 200 square mile (518 km²) area at the border between Montana and Canada, and in 1974 the gray wolf joined the endangered species list.

In 1995 and 1996 the FWS brought small groups of wolves to Yellowstone and central Idaho in an effort to restore an ecosystem that had long missed this predator. Vehement opposition from ranchers that had been building during the 13-year recovery preparation burst into one of wildlife conservation's most rancorous debates. Montana rancher Randy Petrich seemed to speak for many residents near Yellowstone when he declared to the Associated Press in 2007, "I believe that any wolf on any given night, if there happens to be a calf there, they will kill it. In reality, to help us now, we need to be trapping them, shooting them—as many as possible." The gray wolves of the northern Rocky Mountains nonetheless grew to a population exceeding 1,500, and in 2008 the U.S. Interior Department removed it from the endangered species list. Interior Deputy Secretary Lynn Scarlett proclaimed that

Heated controversy swirls around the reintroduction of gray wolves to the Northern Rockies. These wolves in Yellowstone National Park have been praised by environmentalists for restoring ecosystems, and they have been despised by ranchers as a threat to livestock and families. (*Xanterra Parks and Resorts*)

(continues)

(continued)

the wolves "are thriving and no longer require the protection of the Endangered Species Act." The wolves thrived, yes, but lifting the protections would lead to further controversy.

Wolves are a top predator in their ecosystem. Their main cause of death in the wild comes from conflicts with other wolves for control of packs and breeding opportunities. In Yellowstone National Park and in Idaho and Canada, elks are the wolves' main food source. When wolf packs complete their hunting forays, the carcass left behind supplies food for scavengers such as eagles and endangered grizzly bears. Wolves hunt in a supremely efficient manner from identifying their prey to killing it. The lead wolf first observes an elk herd to identify a sick, weak, or old individual, then directs the pack's circling of the herd to disorganize it and separate the prey from the other elk. Finally, the pack's hunting members bring down the individual.

Hunters and hunting outfitters joined the ranchers in condemning the 1995–96 wolf reintroduction because of concerns that the packs would eliminate elk herds within a few years. An Idaho outfitter, Dan Fowler, told the Pacific Northwest's *Spokesman-Review* in 2004, "The wolves are just doing a tremendous amount of damage to our herds. There's not going to be any elk left. It's just a matter of time. It's going to be bad." The wolves have indeed roamed into Idaho, Montana, and Wyoming in a territory covering 113,000 square miles (293,000 km²) and have established about 90 breeding pairs, causing Lynn Scarlett to conclude, "The wolf's recovery in the northern Rocky Mountains is a conservation success story." But the successful reintroduction has an ironic twist that may please sportsmen like Dan Fowler: Hunters may now legally stalk wolves as they did 100 years ago. "The enduring hostility to wolves still exists," warned Earthjustice attorney Doug Honnold. "We're going to have hundreds of wolves killed under state management. It's a sad day for our wolves." Opinions on the Yellowstone wolves seem to have no middle ground.

Evidence has accumulated on the positive effects wolves bring to their ecosystem. Wolves are selective hunters, meaning they cull only the youngest, oldest, weakest, slowest, or injured animals from a herd and thus build a stronger elk population. The average age of elks killed by wolves (about 14 years) is higher than the entire herd's average age; wolves target the older animals. Elk hunters, by contrast, tend to target the herd's strongest, healthiest animals.

Wolf predation on herbivore elk also affects the vegetation in grazing areas in two ways: they reduce herd density so that the grazers are less likely to remove an entire stand of new growth, and they make elk graze on high slopes and spend less time in flat areas near water. While the elk graze mountain slopes, young willows, aspens, and cottonwoods have again begun to reach maturity in Yellowstone. Birds, beavers, and other animals drawn to riparian areas also rebounded within seven years of the wolf reintroduction.

Estimates on the amount of cattle and sheep killed by wolves vary in this emotionally charged subject. The incidence of predation may be as low as 0.1 percent of grazing cattle; ranchers claim the number is much higher. The FWS's most recent (2007) *Wyoming Wolf Recovery Annual Report* stated that in a six-year period after reintroduction, the total number of cattle taken by wolves increased from about 30 to more than 150, about the same rate as the wolf population's growth rate. The FWS now removes wolves that are chronic cattle predators. Ranches can also help reduce cattle kills through the following practices: locate grazing land far from wolf migration corridors; keep grazing cattle far from wolf dens; and cover or destroy open-air carcass pits. The following nonshooting approaches to preventing wolf predation have also been used with fair success in Europe including Scandinavia:

- fladry (red plastic strips tied to fencing), a method adapted from northern Europe
- combinations of noisemakers, rubber bullets, and electric fences
- radio-activated guard (RAG) boxes—a radio-collared wolf in each pack triggers an alarm in the box, which scares the pack and alerts the rancher
- night pens—paddocks for enclosing herds, especially sheep at night when wolves hunt
- range riders—cowboys on horseback ride at dawn and dusk during the summer months when cow-calf pairs graze in backcountry (remote areas far from the ranch)

Audubon magazine reporter Ted Williams recently wrote, "Now that wolves have been restored to the northern Rockies, all that stands in the way of the biggest success story in the history of the Endangered Species Act is ignorance and superstition." These two obstacles are in fact very big threats to wolves in the wild.

> *There's a whining at the threshold—*
> *There's a scratching at the floor—*
> *To work! To work! In Heaven's name!*
> *The wolf is at the door!*
> —*Charlotte Perkins Gilman, 1893*

(continued from page 86)
Plant reintroduction helps reestablish ecosystems that have been damaged or reduced by pollution, invasion, construction, or natural events such as fire. Like animal reintroduction, plant reintroduction uses only species that are natural to an ecosystem. In many cases, however, the natural plant has become endangered or threatened. Therefore, the second role of plant reintroduction is to reestablish in nature plant species that have become scarce in the biomes where they belong.

Restoration biology comprises a field of study that focuses on fixing degraded ecosystems. Preparation steps for plants that will be introduced into the environment consist of the following steps:

1. seed production
2. seed dispersal in the habitat
3. monitoring seed germination
4. establishment of young plants
5. nurturing young plants to maturity

Tree or plant reintroduction requires monitoring to ensure each species establishes itself in its new habitat and can reproduce there. The establishment period lasts a few weeks in which a plant gains a foothold in the ecosystem. To do this the establishment phase consists of a period in which soil root systems develop, a period of steady growth rate, and then a period of slower growth, called closure, when plants compete for sunlight and soil nutrients. Closure leads to the eventual death of the tree or plant, though this may take many years. During this sequence, meanwhile, the plant life provides nutrients, shade, shelter, and other factors necessary to the ecosystem.

Tropical forests hold most of the world's biodiversity, but deforestation of those biodiversity hotspots has been especially hazardous to ecosystems. Tropical forests have therefore become a priority for worldwide efforts to reintroduce plant life to damaged ecosystems. The following table gives examples of plant reintroduction programs now underway in Asia.

The endangered species list contains four different plant groups: conifers and cycads, ferns, lichens, and flowering plants. Reestablishing an endangered plant in its natural habitat employs one of two approaches: complete reintroduction or enhancement. Reintroduction involves recov-

EXAMPLES OF TROPICAL PLANT REINTRODUCTION PROJECTS		
REINTRODUCTION STRATEGY	**BENEFITS**	**AREAS WHERE USED**
fast-growing industrial plantation trees	alternative wood source to relieve pressure on natural ecosystems	Burma, Palau
ecological reserves	protected tropical ecosystems	Indonesia
agricultural buffer zones	provides farmers with alternatives to forest encroachment	Malaysia
seedling programs	land titles to farmers and credits for raising cash crop trees rather than harvesting natural forests	Philippines
local involvement	local communities take responsibility for managing and protecting their natural forests	Nepal
government forest agencies	strengthens protections for native forests	Sri Lanka
agroforestry	programs to ensure fuel wood is available to communities to reduce dependence on natural forests	India

ering seeds from an endangered plant and establishing seedlings in fields or hothouses. Biologists then return the plant to its native habitat. Enhancement endeavors to help plants that are on the brink of disappearing by adding nutrients to the soil or by introducing the same plant from another location. This second choice gives the population greater genetic diversity and so, one hopes, reestablishes a healthy population.

Today plant reintroduction takes place in every biome and in riparian areas, wetlands, and pollution-damaged sites. Plant reintroduction differs from animal reintroduction because it uses genetic engineering and tissue culture to create stronger, long-lived offspring. Plant scientist Carl Leopold of New York's Cornell University helped Costa Rica reestablish forests that had been nearly eliminated 50 years ago. Leopold pointed out the additional benefits to the local towns because of the program: "By restoring forests we are not only improving the native forests, but we are helping control erosion and helping the quality of life of the local people." For these reasons and for the restoration of ecosystems in addition to forests, plant restoration is a high priority in saving biodiversity.

CONCLUSION

Biodiversity loss can be linked directly to habitat destruction, and invasive species play a significant role in destroying pristine habitats. Invasive plants and animals have become so pervasive throughout the world that people may find it difficult to differentiate native from nonnative species. Invaders do not always mean the end of an ecosystem. Many of the vegetables and fruits grown in the United States were carried here by settlers specifically so they would have food in their new home, and almost all domesticated animals are nonnative species. In nature, thousands of invasive species from microbes to multicellular organisms have infiltrated habitats with little or no known effects. But in hundreds of instances, invasive species destroy ecosystems by eating foods that native animals need, taking over space, and bringing new illnesses to native species.

Invasion offers an important case study on how species compete with each other, but in nature species tend to find alternatives to competition. Perhaps the multiple ways in which species avoid deadly competition results from the evolution of all biota on Earth. Biologists often ask some version of the rhetorical questions, "How did so many spectacularly different species all evolve so that they now share the Earth? Why is a jellyfish a jellyfish and not a ponderosa pine?" Natural invasion must play a role. Species living in invaded habitats adapted to new threats or they went extinct. Species therefore developed camouflage, learned to eat new foods, or moved out of their habitat altogether to find a more hospitable place. There are, after all, many different ways to exist in the presence of an invader but only one way to go extinct.

In the face of natural and unnatural invasion, followed by either adaptation or extinction, biologists have fought back. One method to recover lost habitat and lost species is to capture and nurture the last remaining individuals of a species, and then reintroduce them to their natural environment. This straightforward idea has raised a firestorm of controversy in the high-profile reintroduction of gray wolves to the northern Rocky Mountains, but in general animal and plant reintroduction has had few critics.

Invasive species may be more threatening to biodiversity than habitat loss: habitats can be rebuilt in almost original condition, but some species seem to have found their niche as invaders that cannot be defeated by humans. In this sense, invasive species may grow to be the ultimate weapon that destroys biodiversity unless people take aggressive steps to reclaim habitat for the native species that remain.

URBAN DEVELOPMENT

In the last 200 years the percentage of the U.S. population living in cities increased from 5 percent to almost 80 percent. This increase in urban population has also caused more urban sprawl. Urban sprawl is defined as the growth of human populations in areas that previously had low population density. Such outward spread results when people feel the need for more space, safer neighborhoods, affordable housing, and, ironically, a place where they can live closer to nature. Of course, urban sprawl usually destroys natural habitat by reducing woodlands, grasslands, desert, or wetlands. Urban sprawl also fragments habitat and pollutes it, increases the incidence of wildlife killed on roads, and increases soil erosion, which begins a breakdown of many ecosystems.

The spread of human populations is unlike population growth and decline in the natural world. Animals and plants adjust their birthrate in conjunction with resource availability, but people do not do the same when they build cities higher and farther. If people continue consuming resources with little regard for the future, the environment may reach a point at which it cannot sustain the number of people on Earth. Al Gore wrote in *An Inconvenient Truth,* ". . . death rates and birth rates are going down everywhere in the world, and families, on average, are getting smaller. But even though these hoped-for developments have been taking place more rapidly than anyone would have anticipated a few decades ago, the momentum in world population has built up so powerfully that the 'explosion' is still taking place and continues to transform our relationship to the planet." More than twice as many people live on Earth today than lived here in 1950, when the earliest environmental laws were passed.

Populations with fast growth rates contain the fastest-growing urban centers; China and India provide examples of rapid urbanization. These two countries account for more than one-third of the world's total population growth and, though most of their populations live in rural areas, both have very large rural-to-urban migrations. China and India each will soon reach a milestone that the world achieved in 2007: for the first time ever, more people live in cities than in rural areas.

Developing countries in 2007 had twice the birthrate per 100,000 people (22.1) than developed countries (11.0), and their populations increase at about 1.4 percent a year. Developing countries also possess a disproportionate amount of the world's plant and animal diversity. The economies of developing countries are also growing quickly, signaling increases in urban centers, housing, and roads, which all threaten habitat. Booming economies also cause secondary effects on the environment: expansion of agricultural monoculture, air and water pollution, and noise and light pollution. Despite the economic growth in developing countries, a large number of people still live in poverty in these places—people who feel they must harvest their natural resources for income. This combination of increasing economic growth with widespread poverty has had a crippling effect on biodiversity.

Part of the Biodiversity Treaty addresses the needs of developing nations while at the same time addressing biodiversity loss. The treaty made provisions for compensating developing nations so that their citizens would not be forced to destroy native plants and animals for food or income. Though the treaty may be flawed in some people's minds, it represented an important step by calling for cooperation from more than 150 nations to preserve biodiversity in all parts of the world, not just within national boundaries.

This chapter describes the effects of urbanization on biodiversity, current attempts to correct habitat damage, and the clever ways some species have learned to live with humans.

HUMAN POPULATIONS AND BIODIVERSITY

Human population growth affects biodiversity in two main ways: massive numbers of people change the environment—global warming is the primary example—and urban areas expand into less developed areas.

Expansion into rural land brings with it the hallmarks of city life: roads, congestion, pollution, noise, and wastes.

Humanity affects biodiversity in a general sense because of its impact on the earth and the atmosphere. Developed societies are also industrialized societies with a high level of consumerism, so they exert a larger *ecological footprint* on the planet than nonindustrialized societies. An ecological footprint is the amount of land and water needed to support one person and absorb that person's wastes. The amount of cropland, grazing land, forests, and fishing grounds plus carbon-based fuel and nuclear fuel consumption factors into a single ecological footprint. For instance, a typical U.S. resident has a footprint of about 24 acres (0.1 km^2); a typical resident of India has a footprint of less than 2 acres (0.01 km^2). Subtle things also contribute to ecological footprint because not all industrialized nations have footprints equal to that of the United States. Germany's is about 11; France's and the United Kingdom's are each less than 14. The ecological footprints of various nations are shown in Appendix A.

Carbon footprint is a component of ecological footprint and equals the amount of carbon-based fossil fuels used and the waste made by burning those fuels. Environmental organizations such as the Nature Conservancy calculate carbon footprint as follows: the amount of forestland needed to remove from the atmosphere the end-products of burning a unit of fossil fuel. The Nature Conservancy's online calculator determines the carbon footprint for any household by answering questions on energy use, vehicle type and daily travel, diet, and waste recycling. The calculator then automatically determines an individual's or household's carbon footprint as tons of carbon dioxide equivalents produced per year. *Equivalents* means that all greenhouse gases have been converted to an equivalent amount of carbon dioxide based on the degree of global warming they cause. Ecologists have determined that carbon footprint makes up almost half of a typical ecological footprint, and it has grown faster than any other component, especially in countries with a high level of consumerism. For example, the world carbon footprint is 2.64 acres (0.01 km^2) per person, ranging from nonindustrialized countries such as Peru and Somalia that produce a carbon footprint of 0 acres compared with that in the United States, 13.6 acres (0.05 km^2) per person. The United States has the third-highest carbon footprint, behind the United Arab Emirates (22 acres [0.09 km^2]) and Kuwait (16 acres [0.06 km^2]). Watching television, buying clothes, visiting a gym, and spending a day at the beach provide a short list of examples of carbon-emitting activities. Euan Murray, strategy manager of the Carbon Trust in Great Britain, told Britain's *Independent* in 2006, "This piece of work [calculating carbon footprint] is about making people aware that everything they do involves carbon emissions and not just flights and heating their homes." The term *carbon footprint* shows up increasingly in the media, but as Murray suggests, how does a large carbon footprint affect ecology?

The continued growth of ecological and carbon footprints puts animals and plants in peril because Earth's biota can no longer support the current number of people. Since about 1988 the ecological footprint of Earth's human population has exceeded the planet's capacity to support it. This attribute is called *biocapacity,* and as humans continue to stress biocapacity, other species will lose hold on their natural habitats. This scenario may be altered, however, if people make major changes in their relationship to natural resources. The purpose of green technologies is to find ways to preserve biocapacity by conserving natural resources.

Almost every human activity affects biodiversity in a negative way. All types of industries, agriculture, recreation, and general consumption lead in some way to ecosystem degradation. Furthermore, the natural world seems to have become a foreign concept to many. Each year people are injured or killed because they did not understand animal behavior in nature. It is not unusual, for example, to see visitors at national parks leap from their cars and rush too close to bears, elk, moose, coyotes, and other wildlife. *National Geographic* writer John G. Mitchell once observed while visiting Tennessee's Great Smokey Mountains National Park, "Enthralled visitors often assume—mistakenly—that roadside creatures are too tame to be dangerous." Urbanization has simply disconnected people's lives from nature. Many city residents go weeks, months, or even years without experiencing a day free from human influence.

> The natural world is everywhere disappearing before our eyes—cut to pieces, mowed down, plowed under, gobbled up, replaced by human artifacts.
>
> —Edward O. Wilson

Climate change, or specifically global warming, symbolizes the dangers wrought on the environment by human activities. Earth has always possessed natural rhythms consisting of long periods of colder temperatures and long periods of warmer temperatures. This has led some people to believe that global warming results from a natural cycling of temperature rather than a continuous and increasing phenomenon. Kevin Trenberth, a climate scientist at the National Center for Atmospheric Research in Boulder, Colorado, told the *New York Times* in 2008, "Too many think global warming means monotonic relentless warming everywhere year after year. It does not happen that way." Today's combination of greenhouse gases in the atmosphere and deforestation has caused, nevertheless, a dramatic rise in the average of year-to-year global temperatures than at any other time in Earth's history. Greenhouse gases—volatile organic compounds, ozone, methane, and carbon dioxide and other exhaust emissions—trap the Sun's heat in the atmosphere like a glass greenhouse

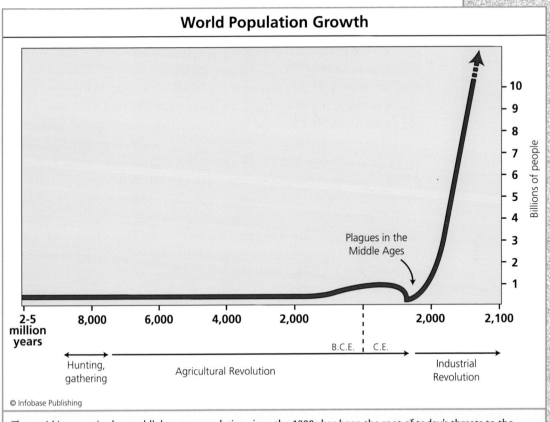

World Population Growth

© Infobase Publishing

The rapid increase in the world's human population since the 1800s has been the root of today's threats to the environment and to biodiversity.

holds heat. At the same time, deforestation leaves fewer trees to absorb carbon dioxide, the most abundant greenhouse gas. In 2007 the Intergovernmental Panel on Climate Change released a report, "Climate Change 2007: Synthesis Report," that was the product of six years of work plus research on previous studies by hundreds of researchers from more than 100 nations. It stated, "Warming of the climate system is unequivocal, as is now evident from observations of increases of global average air and ocean temperatures, widespread melting of snow and ice and rising average global sea level." Put another way, there should no longer be any question of whether climate change is real and caused by humans.

Much of the damage to the Earth's biomes from global warming may be irreversible, but two technologies exist for rescuing the remains of habitats in peril: *restoration ecology* and *reconciliation ecology*. Restoration ecology comprises the activities for returning a damaged habitat disturbed to its original state. Reconciliation ecology comprises the actions that make a habitat suitable for sharing between humans and native species. The case study "The Everglades" discusses how these methods work.

CASE STUDY: THE EVERGLADES

In the past few decades, Florida's Everglades National Park has become a test site for both restoration ecology and reconciliation ecology. The Everglades covers more than 1.5 million acres (6,070 km²) of the southwestern tip of Florida. Hunters, fisherman, and small settlements began entering this forbidding swampland in the early 1800s. For the next 100 years the Everglades suffered almost every type of human incursion: hunting, animal souvenir businesses, burning land for agriculture, logging, aquaculture, construction and development, and tourism. At the same time, burgeoning urban centers siphoned off the Everglades' natural water flow with levees and canals. Neighboring communities commandeered small portions of "the 'Glades" piece by piece.

The Everglades became a national park in 1947 as an attempt to halt its destruction and preserve its wildlife. Even so, people had not yet come to understand the importance of wetlands and swamp ecosystems. Writer Marjory Stoneham Douglas nicknamed the Everglades "the river of grass." This choice of words produced an image of a homogeneous area with little plant variety, but the Everglades represents not a single ecosystem but a blend of related ecosystems.

(continues)

(continued)

The Everglades ecosystem in Florida is one of the world's unique wetlands because of the progression of vegetation types from its northern perimeter to the tributaries that empty into the Atlantic Ocean and the Gulf of Mexico. The Everglades is also one of the biodiversity hotspots of the United States. *(South Florida Water Management District)*

In the park's north, shallow-water sawgrasses dominate, the western section contains a vast cypress swamp protected in a separate area, the Big Cypress National Reserve. Salt marshes and mangrove swamps dominate the southern portion. Salt marshes are ecosystems in which water levels fluctuate with the tides and are the only places where salt-tolerant plants dominate the habitat. Mangrove swamps are ecosystems of brackish water and predominated by mangrove trees, which have the unique ability among plants to absorb carbon dioxide for photosynthesis and oxygen for root function. The Everglades' mangrove swamps provide habitat for crocodiles, alligators, tropical birds, and mammals, and breeding grounds for fish and shellfish. The entire Everglades holds close to 70 endangered species.

On December 11, 2000, President Bill Clinton signed into law the Everglades Restoration Act, making the Everglades the nation's most important ecological restoration project. As part of the restoration, natural resource specialists deliberately alter hundreds of acres of habitat to restore as much of the park as possible. Upon Clinton's signing of the act, the White House issued the following statement: "President Clinton today launched a historic restoration of the Florida Everglades, aimed at reviving millions of acres of sawgrass prairies, mangrove and cypress swamps, hardwood hammocks, and coral reefs. . . . The Comprehensive Everglades Restoration Plan will return a natural flow of water through the Everglades, which has seen more than 70 percent of its historic flows diverted to supply water to farms and communities and roughly half of its acres lost to agriculture and development." This milestone represented a desperate final attempt to recover the complex ecosystem.

The Everglades restoration process begins with tearing down levees, filling canals, and flooding areas that had been leveled for future development. Scientists and engineers also recapture freshwater flows that now go to the ocean and redirect them to areas that had long ago been drained dry. The main techniques for altering flow are surface and belowground reservoirs, freshwater pumping into dried-out areas, and reuse of wastewater. Finally, natural-resource scientists restore wetlands by regulating the amount and frequency of water flowing in other parts of the park that feed the wetlands.

An environmental project as big as the Everglades will surely create controversy. Some environmentalists feel the restoration focuses too much on huge reservoirs and not enough on plants and the earth. In 2007 David Reiner spoke in behalf of the environmental group Friends of the Everglades when he told the *St. Petersburg Times*, "Every year they're falling farther and farther behind, and the only projects being funded are water supply projects." The newspaper's correspondent Craig Pittman added his concerns: "No one has ever attempted putting that many aquifer storage and recovery wells in one place in South Florida, and scientists say no one knows what could happen." No restoration has ever been as big as the Everglades project, so the unknowns seem daunting.

The reservoir building has now stalled because of debates between engineers and environmentalists. No one argues that the Everglades ecosystem needs the massive amounts of water that once flowed through it, but the allocation of the reservoir water has become as controversial as the construction itself, because the water is needed in many ways. The new water flow can go into large reservoirs, constant flows to the south's wetlands, or for the agriculture and populations that have become dependent on the water. The debates often stall the project, leading to more frustration. For example, the National Resources Defense Council's attorney Brad Sewell said in 2008 to the Associated Press, "It makes no sense for them to stop this reservoir because of our litigation. We have never tried to stop this reservoir. Everyone agrees that the Everglades desperately needs more storage to provide more water flows." Meanwhile, tourists would like to continue viewing wildlife in the marshes and scientists want space for ecosystem studies; these activities represent reconciliation ecology rather than restoration.

The Everglades restoration may last another 20 years, and there is no guarantee that the park's critically endangered plants and animals will recover. This area's unique ecosystems are too valuable, however, to allow them to slip away without intensive efforts to turn the tide. The Everglades may in time become a model for restoration ecology, but for now it represents the difficulties of reversing environmental damage.

HABITAT LOSS AND FRAGMENTATION

Habitat is the environment, the physical place, where an organism normally lives. Put another way, a habitat is a species' home. The following are examples of habitats: the Amazon jungle canopy, the African savanna, a North American coastline, an old-growth forest in the Pacific Northwest, and a salt marsh along the Chesapeake Bay. Urbanization has become a major cause of two of the biggest dangers to habitats like these: loss and fragmentation.

Habitat is lost, sometimes irreversibly, for a variety of reasons: development, urban expansion, destruction of agriculture, harvesting native plants, deforestation, invasion by exotic species, or pollution. Before the public became aware of the importance of habitat, many communities eliminated habitats thinking they had helped make the environment healthier. Engineers drained swamps to eliminate disease, leveled woodlands to reduce fire hazard, dammed rivers to store water, and landscaped riparian areas as flood control. All of these activities once thought to be environmental improvements are now known to have contributed to the loss or near loss of thousands of plants and animals.

Habitat fragmentation is the breaking up of a large, continuous habitat into smaller, scattered pieces. Animals isolated in small, fragmented habitats may be more vulnerable to predators, disease, and competition with other species. In addition, if the fragments are far apart, many species are not able to traverse the urban areas in between. Small subpopulations then remain in isolated pockets, and in each one the genetic diversity decreases. Part of today's ecological activities involves ways to save habitats, even in places where urban communities surround them and divide them.

American naturalist Aldo Leopold proposed the idea of restoration ecology during his career with the U.S. Forest Service in the early 1900s. Throughout his career Leopold retained a special interest in techniques for managing and restoring wildlife populations. In 1933 he published his

(opposite) Fragmented habitat leads to rapid decline in wildlife species for three main reasons: loss of feeding areas, reduced land for shelter and raising young, and decrease in the number of breeding individuals to maintain a diverse gene pool. Indian tigers, for example, are a group of tiger species that had an original range from India along Asia's Pacific rim to northern China as well as other sections of Central Asia. Their range today outside of protected areas has been severely limited.

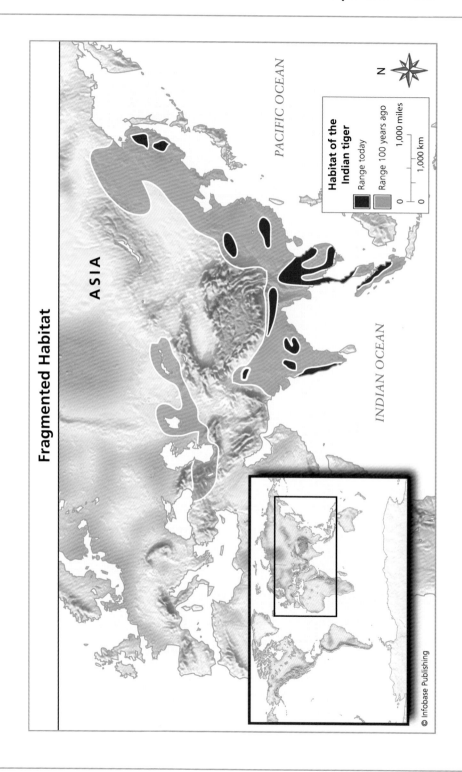

Fragmented Habitat

ASIA

PACIFIC OCEAN

INDIAN OCEAN

Habitat of the Indian tiger

Range today

Range 100 years ago

0   1,000 miles

0   1,000 km

N

© Infobase Publishing

theories on saving habitat in his book *Game Management,* which remains today a resource in restoration ecology. In another book by Leopold, 1972's *Round River,* he wrote, "The government tells us we need flood control and comes to straighten the creek in our pasture. The engineer on the job tells us the creek is now able to carry off more flood water, but in the process we lost our old willows where the cows switched flies in the noon shade, and where the owl hooted on a winter night. We lost our little marshy spot where the fringed gentians bloomed. Some engineers are beginning to have a feeling in their bones that the meanderings of a creek not only improve the landscape but are a necessary part of the hydrologic functioning." Leopold's words describe the evolution of civil engineering into today's environmental engineering, which focuses on construction projects that work with nature rather than against it.

Today's restoration ecologists begin a project by reviewing the living and nonliving components of a region to be restored. They then compile a thorough outline of the species natural to the area by using field studies data and define habitats with input from soil scientists, geologists, and climatologists.

Restoration often begins with fertilization of the soil to replace nutrients, followed by landscaping to induce either water drainage or water retention in ponds. Workers then plant native vegetation selected especially for the habitat. At the final phase, restoration teams leave the restored area and allow it to undergo natural *ecological succession.* This process consists of a sequence of changes in a community over time, the changes made up of various plant life and animal species that thrive in the habitat as it develops. For example, plant successions begin with bare rock and end with large, slow-growth trees. Between the start and the end, the succession progresses from small ground plants, then larger plants, then bushes, and then small, fast-growing trees. A natural succession left on its own restores ecosystems or habitats to almost their original condition.

Reconciliation ecology involves ways to share the Earth with other species even as humans continue to dominate in the habitat. Reconciliation ecology may also be referred to as habitat rehabilitation because some habitats in populated areas can no longer be returned to their original condition, but with good planning they can be made more available to animals. Reconciliation ecologists have their best chance of success when they start at the neighborhood level to save habitat, rather than trying to change an entire urban area. University of Arizona professor of ecology Michael Rosenzweig explained to the campus *UA News,* "Traditional conservation, which sets aside land, is a valuable practice that needs to continue, but there are limits to what it can do." Examples of methods for returning habitat to nature through reconciliation ecology are the following:

- planting bushes and trees that attract birds and butterflies
- constructing birdhouses and bat boxes
- landscaping yards to provide food and a water source for indigenous mammals
- replacing lawns with natural, local plant species
- relandscaping golf courses and cemeteries to contain biologically diverse vegetation
- leaving dead trees in woodlands to provide nesting cavities
- planning urban parks around riparian areas
- constructing migration corridors near or though urban areas

People can protect habitats far from where they live by the decisions they make at home. For example, consumers protect species when they refuse to buy furs, ivory products, shark's fin products, or relics from endangered species. Consumers should also refuse to purchase exotic birds, reptiles, turtles, tropical fish, and animals caught in the jungle, and they should avoid buying rare cacti, orchids, and exotic plants and trees, or wood and paper products derived from old-growth forests.

Ecological restoration takes much longer than reconciliation because it depends almost entirely on natural processes to rebuild components of the habitat. Reconciliation is faster because it alters the environment to mimic a habitat rather than rebuild it. Sometimes successful restoration or reconciliation becomes impossible because the habitat is too badly ruined. In those cases environmental scientists turn to either ecosystem replacement or artificial ecosystems.

In ecosystem (or habitat) replacement, biologists replace a degraded ecosystem with another type of ecosystem that has a greater chance of sustaining native species. For example, a growing town may have destroyed a forest 50 years ago, but the same town can rehabilitate the area by creating a botanical park containing native vegetation. Artificial ecosystems consist of new constructed habitat rather than any attempt to restore the habitat in cooperation with nature. A rebuilt wetland or a rebuilt coral reef offers an example of an artificial ecosystem (or habitat). Though artificial ecosystems sometimes bear little resemblance to the original habitat, many have proved to be successful places for native species to live. Coral reef scientist Thomas Goreau has explained the role of underwater metal structures as a substitute for damaged coral reefs: "Under these conditions, traditional [revival] methods fail. Our method is the only one that speeds coral growth." But the Nature Conservancy's coral reef expert Rod Salm shared with the Associated Press in 2007 his opposing view on the limited value of artificial reefs: "The extent of [coral] bleaching . . . is just too big. The scale is enormous and the cost prohibitive." Perhaps artificial habitat will serve as one of many ways to help species from going extinct, but science cannot depend on any single method to save endangered animals.

SUCCESSION

Habitat restoration requires a vibrant ecological succession to replace the natural plants and animals that once lived in a region. Ecological succession, or natural succession, consists of two components: primary succession and secondary succession. In primary succession, life establishes itself on an area that has little or no living things. Examples of primary succession are plants that grow on rock exposed by a receding glacier or on a cooled lava bed in the shadow of a volcano. Secondary succession comprises a series of related but different living communities replacing each other in sequence. For example, many valleys are actually prehistoric lakebeds that filled with eroded soils over thousands of years and interfered with the lake's aquatic plants. As the plants died, organic debris settled and decomposed and the lake became shallower, then the lake's water evaporated and minerals formed a hard, flat surface. Finally, new bushes and weeds scratched out a living on the bleak, open plain. At some point in this progression, reptiles find this habitat to their liking, predators move in, and as new plants grow, additional species find food and habitat. As a result, a lush valley has replaced an ancient aquatic ecosystem.

Human-made structures also go through succession. These processes may take a hundred years or more but, in the end, a fortunate few places return to the conditions of an earlier time. The following hypothetical

Ecological Succession

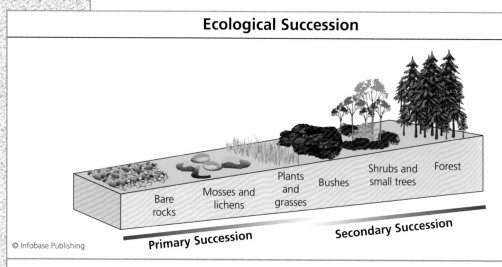

Bare rocks — Mosses and lichens — Plants and grasses — Bushes — Shrubs and small trees — Forest

Primary Succession   **Secondary Succession**

© Infobase Publishing

Ecological succession in plant life creates new ecosystems that develop over decades to centuries. As these ecosystems evolve from simple mosses to multispecies forests, animal populations also change and biodiversity develops. By returning land to nature, biodiversity eventually returns minus the species that have already gone extinct.

example describes situations in which plant and animal diversity changes and species adapt to the existing environment from one phase to the next.

1. a woodland and meadow are cleared for settlers' homesteads
2. generations of descendants plant crops on the land
3. small farms sell to a large agricultural producer that plants thousands of acres
4. the agricultural company goes bankrupt and landowners buy small pieces to plant
5. the landowners plant orchards
6. the landowners convert the orchards to vineyards
7. the vineyards are abandoned and left unattended
8. plants and trees overgrow the vines
9. a woodland reestablishes on the original land

Actions such as clear-cutting forests, plowing meadows, establishing ranch lands, or planting grasslands with agricultural crops disrupt natural succession. Some activities cause succession to go backward to a previous phase. For instance, a controlled fire that clears part of a forest exposes bare rock and leaves only small ground-covering plants. This means the ecological succession must begin again from that point to rebuild the forest.

Urban activities that repeatedly disrupt ecological successions force the animals living there to find new habitat continually. In time this stress may weaken a species' population to the point of extinction.

SPECIES ADAPTED TO URBAN LIFE

Some species manage to adapt to urbanization even though their habitat is gone. Three different modes of adaptation allow some species to do quite well amid dense human populations. First, structural *adaptive traits* enable an animal to hide from predators. During London's industrial revolution of the 1800s, as an example, resident peppered moths darkened from their normal light gray color so they could hide on soot-covered tree bark. This new coloring represented a structural adaptive trait. Second, species inherit physiological adaptive traits, for instance, black bears that invade campgrounds. Their digestion has adapted to table scraps and fast-food leftovers, not at all the food they have evolved to eat! Third, behavioral adaptive traits are changes in an animal's lifestyle to accommodate humans that have encroached into their environment. Raptors such as hawks, falcons, and ospreys have adapted to building nests on the sides of skyscrapers, under bridges, and atop light poles, and they fledge several new families each season.

In all cases in which species have adapted to urban life, the animals have done so because their genetic makeup enabled them to include one or more of the three adaptive traits: structural, physiological, or behavioral. Almost every animal that has adapted to urban life, furthermore, lives as a generalist rather than a specialist: rats, mice, raccoons, coyotes (now living in New York City's Central Park!), and many birds such as crows, starlings, house finches, and house sparrows.

Adaptation plays a critical part in species survival because animals have only three ways to respond to a change in their environment: adapt to the new conditions through natural selection, migrate to another area, or go extinct. The species most likely to go extinct are those that either cannot adapt quickly to changes brought by urbanization or cannot migrate to a new, safer place. Even migrating animals sometimes find natural migration routes blocked by human activities and are then left only with urban wildlife protected areas such as preserves.

Aldo Leopold observed as early as 1933 the capacity of animals to adapt to human incursion into their habitat: "Lewis and Clarke found elk, deer, grizzly bear, and mountain sheep on the flat plains of Nebraska, a country very different from the mountain forests with which we now associate these species." Fortunately for the species noted by these naturalists, the animals had been given enough time over generations to learn how to find food and live at a different altitude than the Great Plains. Urbanization now occurs so fast, perhaps only human-made migration corridors can aid wildlife.

MAINTAINING MIGRATION CORRIDORS

Migrations take place in the air, on land, and in the ocean. Of the 900 or so bird species living in North America, about 75 percent do some sort of migration; worldwide more than 1,000 species migrate. Several sandpiper species travel more than 8,000 miles (12,875 km) from the Arctic to the southern tip of South America, but the champion may be the Arctic tern, which flies more than 11,000 miles (17,703 km) from the Arctic Circle to the Antarctic Circle each breeding season, almost half Earth's circumference. Other migrations extend a mere 500 miles; some songbirds migrate no more than a few miles or less each season. Migrating birds and any other species that must cover distance to find food, breed, or raise young contend with three main threats to life: exhaustion, lack of food, and predators. Ocean migrations cover equally impressive distances. Gray whale migrations are the longest distance among mammals. The whales migrate between the Chukchi Sea off northern Alaska and Baja California in Mexico, about 12,000 miles (19,315 km).

Animal migrations have followed the same paths for centuries. Wildlife biologists do not know all the factors that influence the timing and the routes programmed into the genes of each migrating species, or how generations know to follow the exact same routes. In many cases, however, migrations take place for very practical reasons. African wildebeests, for example, migrate in herds of more than one million animals every year in a constant search for fresh grazing land after their enormous numbers have depleted the local vegetation. Their migration follows a clockwise route more than 1,800 miles (2,900 km) over the Serengeti in Tanzania to Kenya's Mara River. As the wildebeests travel, predator lions, cheetahs, and hyenas sustain their young by taking the herd's unhealthy members.

Migration corridors like those of the wildebeest and the gray whale consist of a strip of land or sea through which the wildlife moves. Each migration corridor must contain a minimum amount of territory to provide sufficient food and protection from weather and predators. Because urban areas have encroached on the lands that create corridors, many migration routes now receive legal protection. In the United States, the Migratory Bird Treaty Act of 1918 bans the killing of migratory birds when they are flying or during their stopovers. The law protects grasslands and plains, coasts, river corridors, and forested areas that the birds use along the migration route, called a flyway. The Endangered Species Act (1973) and the Marine Mammal Protection Act (1972) also outlaw the hunting of migrating animals. International cooperation provides a necessary help toward protecting migration corridors. For example, the International Ramsar Convention (Convention on Wetlands of International Importance Especially as Waterfowl Habitats) of 1972 sets migratory bird treaties among the United States, Canada, and Japan. The United States also cooperates with Mexico for bird and mammal migration protection under the same law.

In the United States, government agencies and wildlife organizations have established joint programs for maintaining corridors. As an example, the Yellowstone to Yukon Conservation Initiative includes the work of more than 800 organizations in Canada and the United States to maintain a migration and habitat corridor extending from the Peel River in Canada's northern Yukon Territory to the Wind River Valley in Wyoming. This corridor supports the migration patterns and breeding of grizzly bears, black bears, elk, wolves, mountain lions, plus many bird and fish species. The Yellowstone to Yukon Conservation Initiative explains the rationale for developing the corridor: "The Yellowstone to Yukon region is one of the last places left in the lower forty-eight [states] where the full historical suite of carnivores and ungulates—including grizzly bears and caribou—can still be found." It goes on to say, "Addressing conservation on the large-landscape scale is revolutionary and timely. It's no longer enough to preserve isolated forests, valleys, and wilderness areas. Connection of habitats is key to the long-term health of ecosystems and the biological diversity that supports both wildlife and human communities." There may be no better definition of ecology.

Urban habitat corridors serve similar purposes as they do in the natural environment. Migration corridors through urban areas help wildlife with

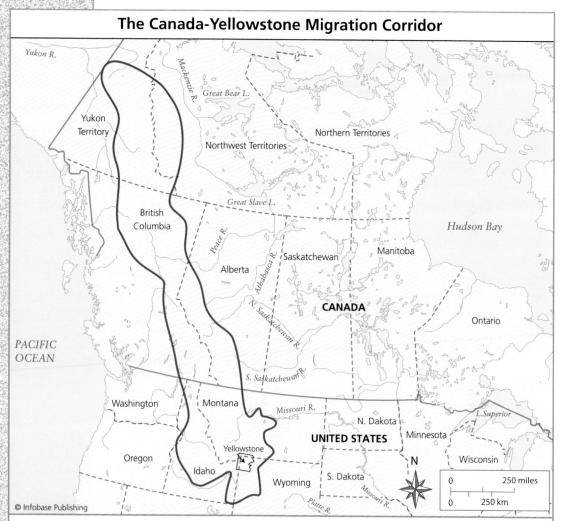

The Canada-Yellowstone Migration Corridor

The migration corridor that extends from the Canadian Rocky Mountains to Yellowstone National Park is called an ecoregion because it acts as an immense ecosystem spanning different climates and elevations and thousands of square miles. This migration corridor has been important in the reestablishment of grizzly bear and gray wolf populations.

the following: connecting fragmented habitats; escaping from floods and fires; escaping from predators; seeking new food sources; providing options for shelter, dens, and nests; providing rest stops for migrating birds; and avoiding construction, traffic, noise, and harassment from humans. The sidebar "Urban Greenways," on page 113, elaborates on the roles of habitat corridors in cities. Traffic from urban sprawl has taken a toll on local biodiversity. In 2007 three rare Florida panthers died on roadways in a single week, because traffic does not discriminate between rare or abundant species, and these accidents kill mammals, birds, amphibians, reptiles, and even marine mammals that climb onto land and enter roadways.

Urban habitat corridors are usually of two kinds: riparian corridors along streams and rivers, and hard-surface corridors, hedgerows, tree lines, landscaping, or fencing. Sometimes people unintentionally build migration corridors, as they do when they plow roads in winter in Yellowstone National Park. Plowed roads allow greater access for snowcraft tours, snowmobiles, and snowshoeing, but the roads also enable the park's bison to roam farther in search of scarce food. The roads have helped weak animals because they no longer must expend energy pushing through snow

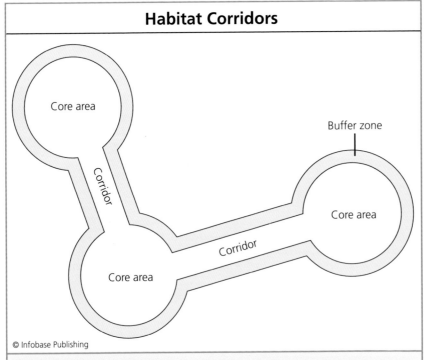

Habitat Corridors

Core area

Buffer zone

Corridor

Core area

Corridor

Core area

© Infobase Publishing

Habitats threatened by fragmentation can be helped by preserving corridors between them. Habitat corridors allow isolated populations to find food and interbreed. The best habitat corridors are protected by buffer zones that contain little or no human interference.

URBAN GREENWAYS

n urban greenway is a band of land within an urban area that serves as open space for nature or recreation. Urban biking and running trails may be thought of as greenways that preserve a portion of urban land for recreation yet have minimal impact on the environment. Urban wildlife greenways are also called biological, dispersal, or ecological corridors (eco-corridors).

The idea of setting aside a portion of an urban landscape for nature has been practiced since the late 19th century. Most of the earliest open spaces consisted of city parks, paths connecting the parks, and botanical gardens. The notion of building open space primarily for animals rather than people took root during the 1960s ecological movement. A few towns managed to carve out space for nature, but these places rarely included wildlife management. The science of melding wildlife habitat, animal lifestyle, and human behavior had yet to gain many followers. In 1986 American conservationist Lawrence R. Jahn wrote, "Some people who now strive to advance conservation programs are prone to forget that there was a time when there was no field of conservation endeavor. Wild things were something to be overcome and forgotten, not something to be cherished, respected, and managed purposefully to ensure their perpetuation and associated values." In a small way, urban greenways bring back those values.

Greenways built in even the most bustling cities serve one of two similar purposes: they provide both people and animals with a natural refuge within an urban area, or they provide principally wildlife with refuge and a route for traveling between habitats. Greenways that fulfill the first purpose provide scenic areas, places for recreation, and spots to observe and enjoy nature. They also add plant life to an area that may be lacking photosynthetic trees, bushes, and gardens. Therefore urban greenways help reduce carbon dioxide buildup in the atmosphere. Mixed-use greenways provide space for both humans and for wildlife and usually consist of narrow strips of land that benefit small birds and small animals.

Greenways as wildlife migration corridors—the second purpose—have several functions in wildlife management. First, some greenways provide habitat for nonmigrating species. Second, they provide hunting

(continues)

(continued)

grounds for predators such as hawks, owls, and foxes. Third, greenways are the main means by which large and small animals travel between habitats for seasonal breeding, finding new food sources, and escaping predators. Fourth, greenways help species that have a genetic need to follow a set migration pattern. Wildlife greenways in fact benefit biodiversity by presenting a species with the opportunity to disperse its genes and giving endangered species a chance to escape threats.

Greenway design has become part of a new scientific field that has emerged in the past few decades. German geographer Carl Troll first used landscape ecology in 1939 to describe this specialty. Landscape ecology blends nature's needs and people's needs in the planning of new urban centers. Only since the environmental movement bloomed in Europe and the United States did this field of study gain momentum. Landscape ecology combines the following disciplines: botany, ecology, ecosystem management, wildlife management, natural-resource management, and urban design. Within these broad areas landscape ecologists today must have knowledge of geography, soils, hydrology, climatology, sociology, economy, regional planning, and civil engineering.

A greenway may be an open meadow, stretch of grassy or wooded land, or riparian habitat, and it must provide the key components of protection, food, and clean water. Riparian greenways should be free of pollution, runoff, and erosion, and they should resist invasive species. Greenways should also incorporate a buffer zone around the area, usually made of woody plants and deep grasses, mixed hardwood forest, or coniferous forest. These zones serve two purposes: they provide extra protection and reduce stress for animals walking through an urban area, and they reduce erosion and runoff adjacent to riparian corridors. Finally, ideal greenways limit recreational use by people and pets.

to find a few bites of grass. Once these animals wander outside Yellowstone, ranchers have shot them in fear the animals will transmit disease to their cattle. Migration corridors, therefore, must be built or maintained with a sound plan as to their effect on both humans and wildlife.

Fencing and other barriers prevent wildlife road losses, but at the same time these barriers halt migrations. Wide divided highways such as interstate highways also cause habitat fragmentation. Amy Masching, a conservation biologist at the Denver Zoological Foundation, explained to Audubon magazine in 2007, "Highways prevent species from accessing their historic migration and breeding areas, which reduces their genetic diversity." For these reasons civil engineers have built landscaped bridges and tunnels in the United States, Canada, the Netherlands, and Australia. Structures have also been developed for large and small mammals, marsupials, birds, amphibians, and reptiles.

MONITORING AND MAPPING THE ENVIRONMENT

Future city design should meet the needs of the populace, animal populations, and native plant life. At the same time the design must be compatible with the local landscape and topography, aided by aerial photography and satellite images. City designers confer with ecologists to study the environment and assess the effects of urbanization on it today and tomorrow.

Mapping helps environmental scientists assess the size and dispersal of human populations, as well as the effects of human activities on the earth and the atmosphere. Today global maps describe the following conditions: atmospheric temperatures, ocean temperatures, surface water plant life, forest area and other vegetation, drought areas, population centers, and light intensity. The National Geographic Society, for example, creates accurate maps by combining satellite images with information in databases on populations. The detailed maps show population densities, growth rates, family sizes, life expectancies, population movement patterns (immigration and emigration), and population age and gender.

High-resolution satellite images have been used for several years by law enforcement and census takers for estimating population size. Census workers begin by taking images of small blocks of a city and then counting the individual people in the block. They then compile block after block of data until they have calculated an entire region's population size. Measurements like this make up part of a field of study known as *population dynamics,* which describes the major living and nonliving factors that cause increases, decreases, or other changes in the composition of a species, human or wildlife.

Of course, almost all of the components of human population dynamics impact biodiversity in some way. The growth and migration of human populations is the single biggest factor in pushing animals out of their habitat, destroying habitats, and causing extinction. Chapter 7 describes the technologies used in mapping important components of biodiversity. Maps and images also provide information on the secondary effects of population dynamics on the environment. For instance, scientists use mapping to monitor the following by-products of climate change: coral reef bleaching, algal blooms (an overgrowth of algae in water due to pollution), phytoplankton levels, desertification, topsoil erosion, distribution of nonnative species, glacier regression, and melting of polar ice caps. Online maps generated by satellite imagery enable the public to find all the hazardous waste sites near them by typing in their zip code. Ecologists use this same technology to assess the wildlife populations most threatened by pollution. The Locus Technologies president, Neno Duplancic, explained the future of environmental imaging in a 2008 interview to discuss his company's technology: "Huge progress has been made over the past forty years. Back then, this industry did not exist. Today it makes headline news every day. . . . Our focus is information management. . . . The amount of environmental data has grown exponentially over the past decade and it will continue to rapidly increase with the emergence of real-time sensing and wireless transmission technologies." As Duplancic suggests, environmental science is entering a phase in which hardly an inch of the Earth has not been scrutinized.

CONCLUSION

Urbanization brings with it an abundance of human activities that influence, disrupt, fragment, or destroy habitat, and so threaten biodiversity. Even more troubling, the world's urban population continues to increase and urban centers increasingly spread outward into formerly untouched animal habitat. Pollution, climate change, noise, roads, construction, industry, and countless other disturbances create an intimidating challenge for wildlife. There seem to be very few ways to correct the damage that urbanization inflicts on habitat and biodiversity every day, but environmental science has made encouraging inroads in recent years to undo some of the worst damage that comes from swelling urban centers. New methods in habitat restoration or habitat reconciliation help to rebuild habitat for species that cannot adapt fast enough to urbanization in order to avoid extinction. Innovations such as human-made greenways offer reasonable substitutes for migration routes that have long ago been blocked by buildings and other structures.

Environmental scientists cannot turn back the clock to return habitat to the world's most vulnerable species, but they can make progress toward building habitats that allow wildlife and humans to share a crowded planet. It is too soon to tell whether we will provide enough sanctuary for wildlife to slow biodiversity loss. New technologies in environmental science will at the least give scientists a better chance to identify biodiversity's biggest threats and develop more methods to conserve species. The test will be to develop species-saving technologies before urbanization eats up every last trace of the natural world.

NATURE RESERVES

Environmental scientists have more than one option for preserving living space for endangered species. Habitats that arise as a result of restoration ecology, artificial habitat construction, and reconciliation ecology create new living space for animals that have been forced out of their home territory. Nature reserves extending over large, continuous tracts of undisturbed land provide the best chance for protecting habitat and thus preserving wildlife. Designated land or ocean habitats help both approaches to saving biodiversity: the species approach and the ecosystem approach. When a nature reserve contains land set aside for a particular species, then this reserve is said to be aiding the species approach. For example, Montana's National Bison Range, run by the U.S. Fish and Wildlife Service (FWS), contains many animal species, but it specifically serves as a bison reserve. Nature reserves that follow the ecosystem approach to saving biodiversity protect entire portions of a biome rather than focus on individual species. The Grasslands National Park in Saskatchewan, Canada, for instance, protects an area for the purpose of protecting an entire ecosystem.

Currently only 12 percent of the Earth's land area remains protected in nature reserves, wildlife refuges, wilderness areas, and national parklands. Humans use the other 88 percent. The world population's average ecological footprint shows that people have overextended the amount of space and resources the Earth can sustain. Without major changes in their behavior toward the environment, humans will eventually use up the planet's resources. One needs very little imagination to guess wildlife's fate at the current pace of population growth and urbanization.

This chapter examines various types of land conservation for preserving endangered animals and plants. A section on zoos and aquariums examines the unique roles they play in species preservation. The chapter also describes some of the biggest threats to these designated areas and the innovative programs biologists use for protecting biodiversity.

HISTORY OF WILDLIFE PRESERVATION

Artist George Catlin traveled the American West from 1832 to 1839, and during his travels he painted portraits of native American life and natural settings. Catlin appreciated the raw wildness of the places he visited and wrote of them in a series of letters to the *New York Commercial Advertiser,* which published Catlin's writings in 1880. Catlin the artist also understood ecology. He felt the threat of advancing settlements of whites on the western prairies. One of Catlin's notes contained a poignant vision of the West's future. He observed, "Nature has nowhere presented more beautiful and lovely scenes, than those of the vast prairies of the West; and of man and beast, no nobler specimens than those who inhabit then—the Indian and the buffalo—joint and original tenants of the soil, and fugitives together from the approach of civilized man." Wildlife preservation in its various forms aims to give wildlife a safe place to live, breed, and behave normally, so they do not have to live like "fugitives."

George Catlin became the first naturalist to propose a means for preserving undamaged lands. He called such places "a nation's Park" and proposed it for both animals and Native Americans. But the U.S. government held a far different vision for the West. On May 20, 1862, President Abraham Lincoln signed into law the Homestead Act to encourage the swift settlement of western territories (the plains, grasslands, front range of the Rockies, fertile valleys of the Midwest, and deserts). Said Lincoln, "I think it worthy of consideration, and that the wild lands of the country should be distributed so that every man should have the means and opportunity of benefiting his condition." Families who had exhausted their farmlands in the East moved westward to cultivate unspoiled lands. Between 1862 and 1960, the Homestead Act transferred into private hands about 10 percent of all U.S. lands in the lower 48 states. Homesteaders staked out grasslands, prairies, and even less hospitable deserts and mountain slopes for new farms and ranches.

Many nature lovers sensed the trouble that homesteading might bring. Of these visionaries, only President Theodore Roosevelt held the power to address the rampant depletion of the country's natural resources. He

The 26th President of the United States, Theodore Roosevelt, developed the Square Deal, a four-point program for the nation. The Square Deal consisted of addressing social problems, regulation of big business, control of the railroads, and conservation of natural resources. *(Library of Congress, Prints and Photographs Division)*

argued that America should put its efforts into conservation as enthusiastically as it put them into consumption: "The conservation of natural resources is the fundamental problem. Unless we solve that problem it will avail us little to solve all others." (In this way Roosevelt was remarkably ahead of his time.) In 1903 Roosevelt established the first federal wildlife refuge at Pelican Island, Florida, but overgrazing and cultivation still threatened western lands. Without western reserves intended only for nature, landowners would soon exhaust the vast resources. In 1907 Roosevelt addressed Congress on the need for legislation to preserve pastureland and grazing land before they disappeared. "The reward of foresight for this Nation is great and easily foretold," he said. "But there must be the look ahead, there must be a realization of the fact that to waste, to destroy, our natural resources, to skin and exhaust the land instead of using it so as to increase its usefulness, will result in undermining in the days of our children the very prosperity which we ought by right to hand down to them amplified and developed." In two terms in office Theodore Roosevelt made conservation a national goal and emphasized forest and wildlife preservation. Following his address to Congress, he called to Washington governors, university presidents, scientists, and business owners for the first National Conservation Commission. Roosevelt's new commission became the first government-led agency to set policies on resource conservation.

In 1935 British botanist Arthur Tansley introduced the term ecosystem in his paper "The Use and Abuse of Vegetational Concepts and Terms," published by the Ecological Society of America. Before Tansley explained ecosystems, biologists and the public tended to think of wildlife in nature as individual species. The idea that living and nonliving things interact in the environment had not emerged. Tansley presented the concept that biology intertwined with physical factors in the environment, and he cautioned scientists to remember the influence of land and climate because ". . . plants and animals are components, though not the only components." But in 1935 at the height of the U.S. industrial growth, would the nation's industrial machine care about the welfare of ecosystems?

From the 1930s through the 1950s, new factories, railroads, and ports appeared throughout the country. In the 30 years after Tansley's introduction of the ecosystem, open spaces, grasslands, and farmlands, rapidly disappeared. By 1965 California tried to halt the trend by passing the California Land Conservation Act (commonly known as the Williamson Act). This act became one of the first pieces of legislation for saving open spaces of unspoiled land. The Williamson Act provided tax incentives to farmers and landowners to discourage them from selling their lands to commercial developers, and also allowed local governments to set up 10-year contracts with landowners in which no fewer than 100 acres would be protected.

The environmental movement, inspired by Rachel Carson's books, fulfilled Theodore Roosevelt's vision for natural-resource conservation. Biologist and amateur environmental scientist Rachel Carson published *Silent Spring* in 1962, which alerted the public to the harm people's actions were causing wildlife. Environmentalists drew increasing attention to the plight of wildlife over the following decade and contributed to the passage of the Endangered Species Act of 1973, which implied that all living things—wildlife and plant life—had a right to exist on Earth. The act's original draft did not mention any species by name, which further emphasized its intent to protect all of nature.

The 1970s ushered in a stronger wave of environmentalism; U.S. and international activists voiced their concerns to a variety of governments and, in some rare instances, developed a role in government. The Greenpeace organization founded in 1971 in Vancouver, British Columbia, Canada, embraced a mission of nonviolent resistance to environmental ruin. Though it began as an independent group and many people thought of Greenpeace as a radical group, Greenpeace International now works closely with national governments. Many ecologists now agree with the *Declaration of Interdependence* Greenpeace published in 1976. The declaration's three laws of ecology are as follows:

- first law of ecology: all forms of life are interdependent
- second law of ecology: ecosystem stability depends on its diversity; more complex ecosystems are more stable than ecosystems containing fewer components
- third law of ecology: all resources are finite and there are limits to the growth of all living systems

These laws have become central principles in all of ecology.

THE WORLD'S WILDLIFE PRESERVES

Humans use about half of the 12 percent of the Earth reserved for nature, meaning only 6 percent of the Earth's land area truly serves wildlife and provides a natural habitat. For this reason conservation biologists and organizations like Greenpeace now propose that at least 20 percent of all land area must be returned to nature to halt the current rate of species extinction. This is because the current land mass occupied by wildlife may no longer be able to sustain these populations over the long term. South African economist James Blignaut and U.S.-born ecologist James Aronson, based in France, wrote in 2008 in *Conservation Letters,* "One fascinating characteristic of *natural capital* is that, unlike built capital, it is capable of maintaining itself. However, that ability of natural capital to maintain itself depends on its 'health.' If critical ecological thresholds are crossed, ecosystems may not only lose their capacity to maintain, renew and replenish themselves, but may also begin to deteriorate spontaneously." Food and water and migration corridors have already declined, and less land for wildlife means that competition will increase to a point at which species cannot continue their normal behavior.

International organizations (see Appendix B) have tried to battle land developers who desire undeveloped land for businesses, housing, or recreation areas. These conservation organizations team with like-minded governments, businesses, and private groups to carry out an ecosystem approach to saving biodiversity. Conservation in its strictest sense would set aside as much land as possible for wildlife with no intervention from humans in these areas. The public and many politicians have argued that this approach is too strict and that national parkland belongs to the people. Recreation and business interests should even take precedence over preservation, they say. Gale Norton was secretary of the interior under

President George W. Bush from 2001 to 2006 and held opinions many ecologists found bewildering for a person meant to protect the country's unspoiled habitats. In 2008 Norton told an audience in Colorado her

The giant panda is an example of an endangered species that has been preserved mainly through the establishment of nature reserves. China's 50 panda preserves protect habitat for about 50 percent of the world's remaining giant panda population. This panda is at home in a zoo in Beijing. *(Connor and Kellee Brennan)*

views on "cooperative conservation," meaning a sharing of all wilderness between wildlife and people. She argued, ". . . many environmental laws are unbending and counterproductive, none more so than the Endangered Species Act. [A farmer] can't harvest his trees because it's full of woodpeckers. That produces exactly the wrong incentives." The *New York Times* spoke for many environmentalists in 2006 in an editorial: "In public Ms. Norton spoke winningly of what she called her four C's, 'cooperation, communication and consultation, all in the service of conservation.' But this was little more than comfy language diverting attention from her main agenda, which was to open up Western lands, some of them fragile, to the extractive industries." The extractive industries referred to here are oil, gas, and mining. Establishing more space for wildlife in an increasingly crowded and energy-starved world has become a difficult challenge for conservationists.

Conservation biology's goal of attaining additional land for wildlife includes three complementary approaches. The first approach involves the immediate end to all destruction of the world's biodiversity hotspots. A second approach focuses on halting the current destruction of old-growth forests. Third, conservationists should save rapidly disappearing lake and river ecosystems. Costa Rica has concentrated on the third tactic to become a model for turning ecology into a business. Between 1963 and 1983 most of this Central American country's forests had been cut for ranch land and Costa Rica had one of the world's highest deforestation rates. Ranching contributed to the economy in a small way; but it mainly benefited a few powerful ranching families. In the 1970s Costa Rica's leaders began a program for reserving land for untamed animals and plants. To counter arguments from ranchers, national funds go to paying landowners who refrain from cutting down forests for cattle grazing land.

Costa Rica's conservation represents one of the world's largest ecological restoration projects. Costa Rican environmental teams have linked large reserves with migration corridors and have surrounded the corridors with buffer zones. Local communities use the buffer zones for sustainable agriculture rather than heavy industry, meaning they grow crops and raise livestock only for their needs in a way that reduces their carbon footprint. For example, farmers use biological pest control, organic fertilizers, soil and water conservation, and minimal non-renewable fuels in the buffer zones. Costa Rican communities receive additional income by using the buffer zones for ecotourism. This combination of ecology and economy to restore habitat is called *biocultural restoration*. Ecotourism has in fact become the largest income source in Costa Rica, a yearly billion-dollar business that supports schools and public services.

Countries that previously depended on heavy industry have taken note of Costa Rica's success. Costa Rica has preserved about 80 percent of its biodiversity; this country the size of West Virginia now contains more bird species than all of North America, and ecotourists put Costa Rica at the top of their favorite places to visit. Ecotourism has the potential to give other countries with biodiversity hotspots a long-term income, which industry cannot promise.

SANCTUARIES AND ARTIFICIAL HABITATS

Sanctuaries and human-made habitats represent a proactive approach to land preservation, and they protect either specific wildlife or biodiversity in general, as in Costa Rica. The role of both sanctuaries and artificial habitats is to provide a setting that allows a species to enjoy health and follow their normal behavior.

As the word implies, a sanctuary is a "safe haven." Wildlife sanctuaries usually operate under the guidance of private, nonprofit organizations that rely on donations for the sanctuary's upkeep. Sanctuaries with healthy funding can accept a larger number of species, accommodate more individuals of a single species, or provide larger habitats containing an animal's natural vegetation and terrain. Many sanctuaries have recently begun specializing in animals that have been abused in captivity, injured in the wild, harmed by poaching, or orphaned. Sanctuaries worldwide give home to almost every animal from butterflies to elephants.

Artificial habitats consist of structures that serve one of two purposes: to provide habitat for species in which their habitat is severely threatened, or to provide habitat for certain species whose presence restores an ecosystem. Rebuilt coral reefs offer a good example of the first purpose. Coral reefs are among the most threatened ecosystems on Earth and the most threatened of all marine ecosystems; more than 65 percent of the world's corals may already be dead or near death. Major reasons for the poor status of coral reefs are the following: rising ocean temperatures leading to bleaching, dynamite fishing, cyanide fishing, dredging, eutrophication,

Oil-drilling platforms have been blamed for ruining marine habitats, but they might also support new habitats for plant and animal life. The underwater structures of platforms provide shelter and artificial habitat for plant life, seaweed, invertebrates, and fish. *(Siemens Corporation)*

disease, and physical damage and removal. Marine scientists have devised artificial reefs as substitute ecosystems for the life that depends on these unique structures. Some of the experimental approaches to artificial coral reef habitat are listed below:

- precipitating calcium carbonate by applying an electric charge to the surrounding sea water; causes calcium carbonate to form and bind to existing coral
- concrete structures
- scrap steel from bridge construction and shipbuilding
- scuttled military and other steel-hulled ships
- retired airplanes

Early attempts at building artificial reefs involved dumping old tires or discarded appliances into the ocean in the hope they would provide strong, permanent structures for sea life. Reefs must provide a stable place for attachment by corals, sponges, barnacles, clams, and other invertebrates. Once these foundation species establish themselves on a reef structure, the structure soon attracts small fish that use the reef's nooks and crannies to feed, hide, and rest. Larger predators such as barracuda, grouper, and snapper then move in to complete the food web. Unfortunately, early attempts at building artificial reefs did not offer a habitat that would allow each phase to happen. In 1972 biologists began putting old tires into the ocean a mile (1.6 km) off the Ft. Lauderdale, Florida, coast. Years passed and aquatic species showed little interest in the tires. The tires began breaking loose and drifting away, sometimes landing on natural healthy corals. Ray McAllister, professor of ocean engineering at Florida Atlantic University, admitted in 2007, "The really good idea was to provide habitat for marine critters so we could double or triple marine life in the area. It just didn't work that way. I look back now and see it was a bad idea." In 2010 Army and Navy salvage crews suspended the cleanup until at least the year 2012, leaving thousands of tires that wash ashore during hurricanes.

Oil companies have considered another type of artificial marine habitat: abandoned offshore oil rigs. After all of the oil has been removed from an undersea oilfield, oil companies prefer to leave their empty rigs at sea rather than dismantling them. Many environmentalists agree with this decision, known as the *rigs-to-reefs* method of making artificial habitat. The Coalition for Enhanced Marine Resources described the benefits of the rigs-to-reef projects off California's shore in its 2006 newsletter *Reef Source*: "California stands to benefit from the thriving habitat through increased tourism, diving and recreational fishing opportunities. For the petroleum industry, this program provides an alternative to full platform removal that is economical and less disruptive to the environment. Demolishing these platforms will literally kill 100 miles (161 km) of creatures and habitat." Companies may also try an alternative to rigs-to-reefs in which demolition workers chop up the steel platform and sink the structure to the ocean bottom, where it then serves as a building block for coral attachment. Oceanographer Sylvia Earle has reported (to the *Orange County Register* in California in 2007) from her dives near the rigs, "You've got a cross-section of life in the sea.

Its diversity is greater than most rain forests." But Linda Krop, counsel for the Environmental Defense Center, countered, "We see birds on telephone polls, seagulls at landfills. . . . Just because there are animals there, that doesn't mean it's a good habitat." Plainly, rigs-to-reefs holds potential promise for further study but, like many conservation plans, it will produce a variety of opinions.

Artificial habitats on land also restore ecosystems by remaking conditions as close as possible to the original habitat. For example, constructed wetlands act as artificial habitats when newly planted natural vegetation establishes a slow and natural flow of water. If constructed properly, the mature wetland soon attracts fish, aquatic invertebrates, migrating birds, ducks, raccoons and fox, and rodents, and supports the regrowth of more plants.

ZOOS AND AQUARIUMS

Zoos and aquariums built before the 1970s obtained their menagerie by removing animals from natural habitats and putting them in confined spaces for the amusement of visitors. Zoos have a long history, possibly gaining the most popularity during the Roman Empire as viewing places for the exotic animals military legions brought back from conquests. Similar collections continued through the Middle Ages until a more modern style of zoo emerged in Europe in the late 1700s. Few people argued the merit of zoos until the birth of the modern environmental movement, when caged animals seemed objectionable and unnatural. Animal advocate Margaret Morin spoke about the issues of traditional-style zoos in 2008 when she discussed an elephant named Jenny at the Dallas Zoo: "She has served her time and now deserves to live the rest of her life at a sanctuary. The zoo is not a humane environment to an elephant." Many zoos have responded to this type of criticism by developing programs for saving the world's endangered species. New York's Bronx Zoo, for example, set up research in breeding endangered and threatened animals. The Bronx Zoo and others have tried to create larger enclosures that provide animals with terrain and plants found in their natural home. Many of today's zoos care for animals in a very different way from metal bars and square concrete cages of older zoos. Full-time veterinarians, animal behaviorists, and biologists work to make zoos as compatible as possible with animals' needs.

Research conducted at zoos often leads the rest of the research world in breeding, techniques for releasing animals into the wild, and wildlife veterinary care. Zoos in several countries

Orchard and vineyard owners have begun to create artificial habitats that control pests without using pesticides. In some instances grassy areas, certain trees, or shrubbery can be planted near the areas to attract birds that eat ground insects and flying insects. Landowners also construct bat houses near crops; bats then eat several different types of moths that prey on crops. The mere presence of bats in an area drives away insects. Small mammals such as rabbits usually enter the cultivated area, and they attract larger predators like hawks, foxes, and coyotes. The predators help control rodents and gophers. In time the orchard or vineyard operates almost as a natural ecosystem.

Zoos and aquariums are a specialized type of sanctuary and artificial habitat in the same location. The sidebar "Zoos and Aquariums" describes how they play a role in wildlife preservation.

run breeding programs that may give endangered species their last best chance of survival. In fact, a number of animals in today's zoos are extinct in the wild and live only in zoos. The International Union for Conservation of Nature's Red List contains a category for these animals: "extinct in the wild."

Aquariums, too, have remade their image as more than amusement parks. Aquariums now offer educational exhibits on marine biology, endangered marine and coastal species, and the effects of pollution on ocean ecosystems. World-renowned aquariums gather data on a variety of topics: species that live in the deepest marine habitats, predator behavior such as that of white sharks, migration patterns of whales, threats to sea otters and other marine mammals, aquatic diseases, and coral reef restoration. Prominent aquariums that combine exhibits with research programs are the Monterey Bay Aquarium (Monterey, California), Shedd Aquarium (Chicago), National Aquarium (Baltimore), and Birch Aquarium (La Jolla, California). Important aquariums outside the United States include the Great Barrier Reef Aquarium (Townsville, Queensland, Australia), Bermuda Aquarium (Bermuda), London Aquarium (London, England), and Centre National de la Mer (Boulogne-sur-Mer, France).

Zoos and aquariums have had a bad reputation based on their past practices. Today zoos and aquariums play an important role in species preservation and education in the areas of conservation and biodiversity.

ANIMAL SPECIES VERSUS HUMAN HUNGER

Ecologists identify poverty as one of the underlying reasons for habitat destruction. Places with high numbers of people living in poverty affect the environment in a variety of negative ways. Areas suffering poverty usually have inadequate sanitation, contaminated drinking water, and poorly managed waste, and the people living in these dire circumstances have crucial concerns in finding water, having enough food to eat, and keeping warm. By trying to fulfill these three basic needs for staying alive, humans threaten habitats and thus biodiversity.

Impoverished areas threaten animal and ecosystem biodiversity in the following ways:

- removal of forests for agriculture
- removal of trees and plants for income
- forest destruction for fuel for heating and cooking
- killing animals for food
- removing animals for exotic animal trade
- poaching endangered animals for animal parts
- overfishing
- destruction of coral reefs for marine animal removal

Environmental ethicists confront the divergent needs of endangered species and impoverished communities. Many ethicists feel the root of the problem lies in an affluent society addicted to consumption, or more specifically, overconsumption. Overconsumption is a behavior in which society buys and uses more resources than it needs for a comfortable life. For example, a crab fisherman working the Chesapeake Bay in Maryland uses a rowboat to do his work, but using a 50-foot (15 m) yacht would be a case of overconsumption. In *WorldChanging: A User's Guide for the 21st Century,* the problem is described thusly: "... we often adapt to our overabundance of choices by picking things haphazardly and acquiring more than we need. The more we own, the more we get used to all the stuff surrounding us, and the less special it feels." Overconsumption can lead to debt, which affects morale, which many people feel can be alleviated by more consumption. As this cycle continues, excess waste pours out and natural resources disappear.

Sustainable lifestyles may provide the best hope for breaking the over-consumption-poverty-habitat loss cycle. New green technologies help consumers reduce their consumption, but a very big part of the solution will be a change in behavior. This rethinking of the standard business model means redirecting economies away from consumerism toward the mission of ending poverty.

POACHING AND WILDLIFE TRADE

Poaching is the illegal hunting of animals on land reserved for their protection. Poaching leads to endangered or extinct species, and in some cases poaching is the sole reason a species has become endangered.

Lone poachers often hunt animals for food or for income by selling them on the illegal exotic animal market. Many endangered animals lose their lives so hunters can remove a single body part: pelts, teeth, bones, whiskers, tails, tusks, or reproductive organs. Lone poachers represent a serious but small facet of the illegal animal trade, but poaching has also grown into a worldwide organized crime. Poachers make lucrative earnings on the black market, the system of buying and selling items that are illegal to sell in conventional markets. Many of the animals or animal parts traded on the black market have become valuable for an ironic reason: laws against the hunting, killing, and selling of endangered and exotic species. The more poaching removes endangered animals from the environment, the more they become endangered. As a result,

President Theodore Roosevelt began his advocacy of wildlife protection as a means to preserve animals as a resource for hunting. Although Roosevelt was an avid hunter, he also respected wildlife for its intrinsic value. After leaving office in 1909, Roosevelt traveled to Africa and South America. He was injured and contracted malaria during an expedition to Brazil's Amazon region in 1913–14, which led to his death in 1919. *(Library of Congress, Prints and Photographs Division)*

the animal's selling price goes up and the animal becomes even more sought by poachers.

The Convention on International Trade in Endangered Species of Wild Flora and Fauna (CITES) prohibits international buying or selling of any endangered plant or animal or parts. Poachers remain bold and vicious hunters, but since the CITES treaty was signed in 1975, it has reduced losses of elephants, crocodiles, and chimpanzees, among other endangered animals. CITES additionally restricts trade in almost 30,000 species classified on the Red List as near threatened. Enforcement of antipoaching laws falls to individual countries that recruit national park officials and hired guards to stand watch over endangered animals.

The Trade Records Analysis of Flora and Fauna in Commerce (TRAFFIC) is an international organization that monitors global trade in wildlife. TRAFFIC also publishes the trends it uncovers in wildlife trade; it was one of the first organizations to alert the world of an enormous market that sells sharks and shark parts. TRAFFIC also monitors trade in endangered plants and trees. International groups like CITES and TRAFFIC will probably never end poaching unless they can enforce stiff penalties against illegal trade. Currently poachers pay only small fines or slip through legal loopholes from one country to another. Meanwhile endangered animals live as targets.

The case study "Protecting Palau's Biodiversity" describes another troubling threat to species: tourists who take animals or plants as souvenirs.

BREEDING AND RELEASE PROGRAMS

Captive breeding encompasses the controlled mating and breeding of animals in wildlife research centers, zoos, and aquariums, also referred to as captive propagation. These programs' primary focus is critically endangered animals and plants, but animal programs present greater challenges. Animal breeding programs include considerations of the reproductive cycles of a mating pair, compatibility between male and female, requirements for birthing and nurturing newborn, and means of safeguarding the mother and youngster from undue stress.

Some captive breeding centers take animals from the wild to add to their program for these two reasons: to rescue an animal from a severely endangered population or a severely damaged habitat, and to increase the

CASE STUDY: PROTECTING PALAU'S BIODIVERSITY

The Republic of Palau is a Pacific Island nation located 500 miles (800 km) east of the Philippines and 500 miles (800 km) north of Papua New Guinea. Palau is famous for its terrestrial and marine biodiversity; it contains one of the largest collections of species found on Earth. Like Costa Rica, Palau has decided its welfare lies in preserving its unique biodiversity and earning its living as an ecotourism destination. A short summary of Palau's staggering biodiversity is as follows:

- 343 islands
- 5,000 species of insects
- 141 species of birds
- 46 species of reptiles and amphibians
- 1,300 species and varieties of plants
- 11 species of dolphins
- 15 species of whales
- 425 species of hard coral
- more than 1,300 species of reef fish
- 111 species listed as threatened on the Red List
- home to seven of the world's nine giant clam species.

In 2003 Palau's governing body passed the Protected Areas Network Act, which establishes most of Palau's undeveloped land as nature reserves. The 20 or so reserves contain restrictions on removal of plant and animal species, and a few areas allow limited access for low-impact activities (observing wildlife, note taking, and photography). Biologists close any areas that begin to show habitat damage, decline in plant growth, or stress on animal life. Palau's residents conduct sustainable farming to eliminate chemicals and erosion, and they raise crops and harvest fish in amounts no more than what they need.

(continues)

(continued)

Palau is a Pacific Island nation home to more than 100 threatened species on the IUCN Red List. Palau owns one of the few remaining healthy reef and fish communities and has one of the largest undisturbed mountain forests in Micronesia. The nation's greatest challenge is protecting these resources further in the face of ecotourism. *(Paul Black)*

More than 60,000 visitors come to the islands each year, mainly for diving and ecotourism, which feeds Palau's economy. The Palau Conservation Society believes the island's ecotourism is also the greatest threat to its biodiversity. Increased numbers of visitors may translate into more hotels, development, dredging, and more people in the reserves. Many tourists scramble over reefs to watch the colorful life under the surface and observe Palau's giant clams. Unfortunately, tourists have also harassed the clams and other species. The Palau Conservation Society's mission focuses on maintaining a conservation ethic on Palau for generations to come. "For Palau, the environment is our economy," President Tommy Remengesau, Jr., said to the Nature Conservancy. "Our people rely on the food and income the reefs provide—and coming generations will, too." Illegal taking of endangered species from one of the last healthy biodiversity hotspots is a tragedy of the same magnitude as organized poaching.

genetic diversity in the group that is already part of the breeding program. Examples of animal populations that have benefited from these programs are the California condor (see chapter 2), red wolf, golden lion tamarin, Arabian oryx, Mongolian wild horse, peregrine falcon, Guam kingfisher, and Hawaiian goose. Aquariums conduct their own captive breeding programs to replenish endangered marine populations. In recent years the Monterey Bay Aquarium has embarked on a plan to be the first to breed a white shark in captivity; in the meanwhile it has housed and released transmitter-tagged white sharks back into the ocean, which the aquarium scientists then monitor via satellite signals.

Release programs work with animals originally captured from the wild and animals born in captivity. Release programs give biologists tense moments; an animal released to the wild is the culmination of thousands of hours of work. Jane Ellen Stevens of the Monterey Aquarium project said in 2008 shortly after a juvenile shark's release, "We're keeping our fingers crossed that he'll make it past fishing nets and hooks to survive to adulthood." The breeding centers increase their chance for a successful release—one in which the animal survives in the wild and joins a social group—with vigorous training before the release. Volunteers and biologists teach animals to avoid humans, dangers such as traffic and power lines, and predators. If possible animals also receive training on proper behavior in social groups. This training responsibility often falls to a mentor animal that lives permanently at the breeding center.

Veterinarians check to assure all animals are healthy; sick or injured animals do not return to the wild until they have recovered. Biologists then take them to the release site. The gray wolves of Yellowstone National Park stayed in pens for two months to acclimate to their new surroundings before biologists released them. Acclimation periods help animals learn about the scents and sounds of the new habitat, daylight patterns, and weather. Each biologist's wish is to watch a healthy animal trot, swim, or fly into the wild without a glance back.

CONCLUSION

Ecologists agree that nature reserves offer the best opportunity to preserve biodiversity. Reserves are critical because only 12 percent or less of the world's land offer legal protections for wildlife; the rest belongs to human activities. Conservation biologists depend on reserves, sanctuaries, and artificial habitats to maintain populations of endangered animal species. The most endangered species may require technologies provided only by zoos, aquariums, and a small number of veterinary colleges. For this purpose zoos and aquariums have transformed themselves from places that once contributed to animal endangerment to important protectors of biodiversity.

Some countries have turned to ecotourism rather than industry for their national income and therefore protect their biodiversity hotspots. But humans remain a threat to protected places, even nature reserves. Poaching and the international wildlife trade remove many animals from protected habitats each year. Regardless of the troubles that some people bring to endangered species, many other dedicated individuals have put substantial efforts into setting aside land for these animals. Protections come in the form of government-run reserves, private sanctuaries, and zoos and aquariums. Even the largest reserves have boundaries, so they cannot be thought of as truly wild. But restricted places like nature reserves may become the only places where critically endangered animals can have a future.

SPECIES PROTECTION

Some of Earth's species would not survive contemporary human activities without special protection. Endangered and threatened species find themselves in vulnerable positions because they cannot withstand the effects of human activities, or they are unable to adapt to changes in their environment fast enough to sustain their population. This chapter looks at various aspects of species protection. It covers the human behaviors and decisions that endanger species and also save species. The chapter examines the options conservations have at their disposal to convince people that nature should be appreciated and understood. The sidebar "Lewis and Clark as Naturalists" on page 142 describes the connections that wildlife and commerce have had in history. This chapter also describes a high-tech approach to the fight against poaching and less technical ways in which people observe nature and bring nature back from the point of disaster. Finally, the chapter reviews the steps taken when an animal has regained its place in the environment so that it need no longer be on the endangered species list.

STEPS TOWARD PROTECTING SPECIES

Species protection and the prevention of biodiversity loss may use focused conservation methods or broad conservation programs. Examples of focused ways to restore habitats consist of artificial habitats, captive breeding programs, and reintroduction programs. National networks of protected nature reserves, by contrast, represent a broad approach to species preservation.

In order to slow biodiversity loss, natural-resource management must be compatible with society's needs. These needs vary. In poor regions,

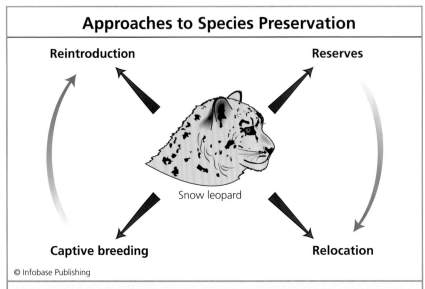

Approaches to Species Preservation

Reintroduction   Reserves

Snow leopard

Captive breeding   Relocation

© Infobase Publishing

More than one wildlife management method preserves critically endangered animal species. Reserves and relocation help restore populations that contain several breeding pairs. Captive breeding programs have been used for the most critically endangered species, with a final goal of reintroducing the individuals into their natural habitat.

people most need water, food, and shelter, but in wealthy industrialized regions residents devour fossil fuel and other energy sources. Protecting biodiversity, therefore, takes into account economics and cultural factors.

The steps toward protecting species fall into several categories. The factors summarized in the following table illustrate the many aspects of species protection.

HOW HUMAN DECISIONS ENDANGER WILDLIFE

Human behavior affects the future of wildlife in various, sometimes unexpected ways. Psychology offers explanations as to why people behave as they do. Psychology may soon play a bigger role in conservation biology, because biodiversity has been victimized by the conditions of too little and too much.

Parts of the world struggle with too little income, poor education, declining health, and limited natural resources to sustain families and

STEPS IN PROTECTING SPECIES	
FACTOR TO EVALUATE	**WHY IT IS IMPORTANT**
biology	critically endangered, endangered, and threatened animals will go extinct before species that are abundant and not threatened
natural resources	habitats are as endangered and threatened as animal species
law	national and international laws exist to protect types of animals, specific animals, and habitat
economics	biodiversity preservation plans have a better chance for success if they do not conflict with profit-making activities
sociology	poverty decreases the effectiveness of biodiversity preservation programs; unchecked population growth threatens biodiversity
culture	some societies have a desire to kill endangered species for cultural or religious reasons
ethics	biodiversity protection requires people to choose between human needs and preventing a species' extinction
politics	species protection usually requires cooperation between friendly and nonfriendly countries

still protect the environment. Those living under these circumstances may decide to clear a habitat to grow food or hunt an endangered animal because of either ignorance or desperation. It is difficult to blame people for harming the environment if they struggle in conditions in which their lives have little margin for error.

Other choices people make are optional and not based on desperation. The sidebar "Environmental Ethics" in chapter 2 describes how a perceived need for an ornamental item, a dagger made of rhinoceros horn, can drive a species toward extinction. Religion also influences the fate of nonhuman species. Some religions embrace the idea that humans have more value to the universe because they are closer to God. As a consequence, their followers might assume that humans have more right to the planet than other animals do. War, too, is fought mainly for religious or political reasons. Some sociologists predict wars will be fought increasingly over natural resources such as water and land, diamonds, and oil. War zones wreak havoc on sensitive habitats. Animal communities of the deserts, forests, wetlands, and oceans have all been in the paths of history's conflicts. "Warfare between modern and traditional societies has often involved what might be termed 'ecological attacks.' One of the most obvious examples of this is the final destruction of the great herds of American bison, the foundation of Plains Indian life in the United States, which closely coincided with the defeat of the Sioux and Cheyenne in the 1870s." Those words were written by Jay Austin and Carl E. Bruch in their book published in 2000, *The Environmental Consequences of War: Legal, Economic, and Scientific Perspectives.* The authors refer to modern societies as those established by European settlers in the New World that would eventually evolve into today's urban centers; traditional societies were the Native American societies based on hunting, fishing, growing crops, and trading. The Plains Indians used American bison in a way that preserved the enormous herds that once covered the western plains, but the westward migration of white settlements wiped out all but a remnant of the American bison population.

Some habitats such as the open ocean and the air serve as *free-access resources*. This means no person owns them and they are common property of all the Earth's species. White settlers likely viewed the open plains and bison and other wildlife there as a free-access resource. Today, people may be less inclined to protect these things because they will not be held accountable. The psychology of free-access goes like this: "No one should care if I dump one load of trash into the ocean because it cannot possibly harm anything as huge as the ocean." Of course, when a million or even a thousand people feel the same, the environment suffers.

Finally, the problem of too much threatens biodiversity. Affluence is the opposite of poverty, and it comes about through the ability to obtain the resources needed to sustain a healthy and constructive life. *Affluenza,* however, refers to an unsustainable desire to overconsume, meaning a person uses more resources and produces more waste than necessary. The large U.S. ecological and carbon footprints provide evidence that the majority of U.S. residents live a life of affluenza. One goal of sustainable living is to break the affluenza cycle and thereby reduce society's ecological footprint. Sustainable living begins a process that leads to the preservation of biodiversity, but the question remains of whether these changes will occur in time, if they occur at all, to save thousands of species from extinction. In truth, the U.S. economy counts on affluenza to keep industry healthy and provide jobs, both of which contribute to the government's income through taxes. It is easy to see why affluenza prevails, even though most people know that overconsumption puts the environment at risk.

GLOBAL CONSERVATION STRATEGIES

Species protection today employs a combination of local efforts by private citizens, called grassroots environmentalism, and far-reaching international laws. International treaties and laws offer the best chance to affect biodiversity loss, because they address entire biomes that stretch across the globe. Two international organizations have had perhaps the greatest effect on protecting endangered species. The first is the World Conservation Union (IUCN), which was established in 1948 by the United Nations. The IUCN manages the Red List and also organizes meetings among its member nations to establish laws banning poaching, exotic animal trade, and habitat-destroying activities. This organization follows the philosophy of conservation mentioned at the beginning of this chapter. That is, it brings together experts in economics, business, social issues, and natural resources to reach agreement on plans for protecting biodiversity. The case study "Lake Victoria" on page 145 describes one of the IUCN's projects that balanced environmental issues with economic needs.

The second important international group is the Convention on Biological Diversity (CBD), which also relies on the cooperation of nations to establish biodiversity programs. It resembles the IUCN because it includes hundreds of government and private agencies that work on biodiversity protection. The CBD focuses on the world's biomes (the ecosystem approach to biodiversity protection) and creates programs for each.

Like the IUCN, the CBD recognizes that economics and politics influence large-scale biodiversity programs. Finally, the CBD gathers all the new technologies available for saving species and habitat.

LEWIS AND CLARK AS NATURALISTS

In 1803 President Thomas Jefferson decreed that army officers Meriwether Lewis and William Clark lead an expedition to the western reaches of America. Their mission: find a water route across the continent for the purpose of opening commerce to the west. President Jefferson's Louisiana Purchase had created the possibility of exploring the country's seemingly limitless natural resources, so the Lewis and Clark expedition set off in 1804 intending to gather information on the land west of the Mississippi, and simultaneously search for a water route to the Pacific Ocean.

Lewis, Clark, and their men, the Corps of Discovery, proved to be astute observers of life on the plains and through the forests. They saw how Native Americans lived in a way that complemented the environment rather than exploited it. In addition Lewis and Clark took notes on the natural resources they saw on their trip. Their records remain arguably the best accounts of this country's natural history. One of many detailed entries in Lewis's journal appeared on January 2, 1806: ". . . the large and small or whistling swan, sand hill Crane, large and small geese, brown and white brant, Cormorant, . . . mallard, Canvasback and several other species of ducks, still remain with us." Much of biologists' knowledge of wildlife and plant life before the westward movement came from Lewis and Clark's notes.

Lewis and Clark's records describe the striking biodiversity that existed on the North American continent at the beginning of the 19th century. By the time the two-year expedition ended, their team would collect specimens and observations from nature that would far outweigh any contributions to commerce. Lewis and Clark also offered an example of how wildlife–monitoring can be done without cameras, satellites, or other electronic gadgets. Observing wildlife in their natural habitats involves a fair amount of intuition about abundance and scarcity. Seeing the same species in large numbers every day provides a good sign that the species is abundant, while spotting an animal only occasionally over a long period is a clue that the species may be scarce. The following table lists some of the hundreds of species seen for the first time by white men and recorded between the Missouri and the Oregon coast. Many of these species are now endangered or have gone extinct.

Lewis, in describing the white cliffs of the Missouri River Breaks in Montana (1805) may have produced one of the most captivating images of the west before white settlements encroached: "The hills and river cliffs which we passed today exhibit a most romantic appearance. As we passed on it seemed as if those scenes of visionary enchantment would never have an end; for it is here

Conservation strategies that require the cooperation of diverse governments have special challenges. National identity, culture, and type of government impact the success of conservation programs because

too that nature presents to the view of the traveler vast ranges of walls of tolerable workmanship, so perfect indeed are those walls that I should have thought that nature had attempted here to rival the human art of masonry had I not recollected that she had first began her work." Lewis and Clark's observations on the living and nonliving features of nature testify to their skill as ecologists before the science of ecology had a name.

SPECIES RECORDED BY THE LEWIS AND CLARK EXPEDITION (1804–1806)	
GROUP	**SPECIES**
plants	more than 24, including wild ginger, bitterroot, Indian tobacco, lady-slipper, wild rose, valerian, Oregon grape, dogbane, Jerusalem artichoke
trees and shrubs	more than 15, including Douglas fir, cottonwood, aspen, whitebark pine, yucca, Osage orange
birds	more than 50, including Carolina parakeet, Clark's nutcracker, sage grouse, spruce grouse, various pheasants, Lewis' woodpecker, meadowlark, Mississippi kite
reptiles	horned lizard, snail, garter snake, rattlesnake
fish	candlefish, salmon, cutthroat trout
large mammals	pronghorn, American bison, Roosevelt elk, grizzly bear, moose, various deer, bighorn sheep, various whales
small mammals	more than 20, including badger, bobcat, coyote, long-tailed weasel, various hare, various squirrels, mountain beaver, wolverine

different societies view species according to different value systems. For instance, some societies kill horses or dogs for food without controversy, but other societies find this unacceptable. A public disagreement between Japan's government and global environmental groups has centered on Japan's yearly hunt in the Southern Ocean (also known as the Antarctic Ocean) for minke and fin whales. Japan values the five-month hunt that kills about 1,000 whales as a part of its culture and a four-century tradition. Some national leaders in countries such as Australia have given their uneasy approval to the hunts, but Israel has taken a strong antiwhaling position. The United States has taken a moderate antiwhaling stance that has fluctuated over the past decade.

Type of government influences the decisions that affect biodiversity. A dictatorship might be a difficult place for an environmental group to make progress in saving biodiversity, especially if the group needs the help of the government to protect land for habitat. Democracies present their own challenges in solving environmental problems. Most democracies provide an open forum for citizens to express different opinions. This can lead to conflicts, such as disagreements between companies that want to drill for oil in the protected Alaskan wilderness and environmentalists who want to preserve the wilderness. Democracies furthermore change administrations periodically so that planning long-term environmental programs can be difficult.

Organizations such as Conservation International serve a role in creating dialogue between national governments but also local governments or political parties so that conflicts do not overshadow the environment's needs. These types of organizations that work free of political influence offer the best hope for keeping biodiversity programs moving forward and making a difference to endangered habitats and species.

SAVING MARINE BIODIVERSITY

The CBD works with many other organizations devoted to biodiversity along the world's coasts and in open waters. International agreements are particularly needed in marine conservation because the ocean is an open space used by many nations.

Marine organisms produce one-third of the oxygen humans breath, yet people tend to forget about the ocean when they think of biodiversity. The ocean has long been treated with disregard because it is a free-access

Case Study: Lake Victoria

Lake Victoria in East Africa covers an area about the size of North America's Lake Huron (26,560 square miles; 68,800 km²). Kenya, Tanzania, and Uganda share the rather shallow (262 feet; 80 m) body, which once contained some of the world's richest freshwater biodiversity. A little more than 50 years ago, a fishing industry grew along Lake Victoria's shores with a booming population. Overfishing during that time reduced the lake's 500 fish species by half; at least 200 of the remaining species teetered on the verge of extinction. Ecologists have labeled it the largest mass extinction of vertebrates in modern times. Sewage and waste poured into the water from towns ringing the lake, leading to eutrophication along the shores. (The lake contained five to 10 times more algae—the cause of eutrophication—in the 1980s than it did in the 1960s.) The pollution accelerated as trees around the lake disappeared because of more development, while nonnative water hyacinth plants spread over the surface and accelerated eutrophication. To make matters worse, in the 1960s the fishing industry introduced nonnative Nile perch to the lake to improve its fish stock. The aggressive predator, weighing in at 155 pounds (70 kg), devoured the

(continues)

This placid view of Africa's Lake Victoria does not give an indication of the lake that now faces threats from shoreline development, pollution, and invasive species. Though the lake requires further cleanup, a cooperative effort among the countries that border the lake and scientific teams have made progress in restoring its ecosystem. *(Michael Shade)*

(continued)

lake's native species and so destroyed many of its food webs. By the late 1990s Lake Victoria had turned into such a toxic pool, even the almost indestructible perch began declining.

Lake Victoria's fishermen soon discovered that perch brought in more income than did most of the native species. Large fisheries were developed that began to haul out perch by the ton and sold them to processing plants. By the 1980s, it had become clear that the lake could not continue its biodiversity decline and sustain the livelihoods of even these large businesses. Local residents who depended on fishing the native species to feed their families had been forced to give up fishing when the native species disappeared. They turned to menial jobs at the perch processing plants that were providing a high-priced product for customers in Europe and the Middle East. If the perch continued to decline in the lake's unhealthy environment, the processing plants and their workers would both succumb.

The IUCN reacted to Lake Victoria's coming crisis in the 1980s and 1990s. It managed to save many native species by placing them in aquariums around the world. Meanwhile plans for Lake Victoria's revival centered on the needs of the local economy versus the future of the lake's ecosystem. Because the death of Lake Victoria would devastate the region's economy and lead to political tension, any recovery had to include economic considerations with ecological ones. One plan called for the control of the Nile perch and the water hyacinth to allow the lake's natural ecosystems to rebound. Unfortunately, a healthy ecosystem restored to Lake Victoria might never generate the money that the perch could. Aquatic reserves seemed to offer better hope for Lake Victoria's misplaced species. One plan called for sheltered inlets in the lake, protected by netting. Native species could be returned to the protected areas while the dominant perch would live in the main part of the lake. A second plan suggested the use of nearby lakes to provide the native

resource. As such, industries and individuals have erroneously assumed that a little of their wastes dumped into the water cannot possibly harm the vast ocean. The fact that people usually cannot observe ocean pollution compounds the problem. Moreover, the marine habitat is not a single, homogeneous area that quickly digests wastes; it is a combination of specialized and sensitive ecosystems.

Marine ecosystems include coastal areas as well as the high seas. The important marine ecosystems include: coral reefs, mangrove swamps, delta marshes, seagrasses, algae and phytoplankton communities, fish with the same conditions they had in Lake Victoria. A third plan targeted the water hyacinth. Removing the hyacinth would improve the lake's oxygen levels and help the perch rebound. This would save the perch-based economy and also buy time to set up protected reserves for the native species, thereby returning to the local residents their preferred way of life.

But how were biologists to remove hundreds of thousands of tons of hyacinth? Australia offered a solution based on its own weed-eradication efforts. Lake Victoria restoration made use of two South American weevils, introduced from Australia, which had voracious appetites for the hyacinth plants that were choking great portions of the lake. Local communities took some convincing that a single bug could save an ecological disaster. Australian biologist Mic Julien recounted in 2003, "... everyone was mesmerized by the notion of a quick fix ... chemicals and big shiny harvesters. All we were offering were tiny weevils." Within two years of the weevils' release into the lake, in 1997 the water had been largely cleared of hyacinth. The weevils were a chemical-free method of eradication. As the hyacinth disappeared, the weevil population also declined, returning the water almost to its original condition.

Today Lake Victoria has become a model of saving an ecosystem that had been all but lost. The lake's ecosystem still has a long way to go, however, to full recovery. Despite the encouraging progress, the Global Nature Fund cited Lake Victoria as 2005's most threatened lake ecosystem, and the fight to continue biological control methods goes on. Julien said, "There were significant gains for people backing the chemicals and machines, but there was no profit in biological control, except for the community at large." The Lake Victoria cleanup offers an example of one peculiar characteristic of today's efforts to save biodiversity: Many people support it only if they can earn money in the process.

pelagic (open) waters, and deep-sea waters. Coastal areas contribute a disproportionate amount to marine biodiversity. For instance, marine biologists point out that though coasts account for only 10 percent of the globe's ocean, they contain 90 percent of all marine species. Moreover, only a small portion of coastal habitat is protected by environmental laws. One such protection in the United States are Marine Protected Areas, marine areas temporarily or permanently closed to harvesting resources. The Woods Hole Oceanographic Institution in Massachusetts described the benefits of protected areas in *Oceanus* magazine in 2008. This entry

U.S. coastal zones have been protected since 1972 by the federal Coastal Zone Management Act, which established a national program for conserving, protecting, and restoring coastal areas. These zones include wetlands and estuaries, seen here in the Everglades, as well as dunes and coral reefs. *(Everglades National Park)*

describes the Georges Bank protected area: "These closures have given us a unique opportunity to examine a marine protected area in a temperate system under a 'macroscope'—to examine how marine ecosystems are structured and how they function and recover." As this article attests, protected marine areas give scientists their best opportunities to study and save species.

Pollution, invasive species, and global warming threaten marine ecosystems as much as they harm land ecosystems. Marine ecosystems also contend with eutrophication and overfishing. Marine research, furthermore, lags behind the science of land ecosystems because deep-sea waters remain largely a mystery; they are difficult to reach and difficult to study. The list below contains some of what we know about the perilous state of marine ecosystems.

- 50 percent of wetlands have been lost to agriculture and urbanization
- 10 percent of beaches have been lost to rising ocean levels and development

- 50 percent of coral reefs are threatened; at least 20 percent have already died
- 40 percent of mangrove swamps have been lost to agriculture and development

Another obstacle to studying and protecting ocean ecosystems is that open waters belong to no single governing body. International organizations now cooperate on almost every aspect of marine biodiversity, and they have solicited the help of governments and universities. Maritime laws also exist to stop ocean pollution, illegal fishing, coral harvesting, and other assaults on ocean ecosystems. These laws are difficult to enforce, however, because of the huge expanse of ocean they cover. Conservation International has pointed out that almost all of the ocean's biodiversity hotspots, called seascapes by this organization, cover large areas inside and outside national ocean boundaries. Of all biomes the oceans may be in the greatest need of international cooperation for preserving biodiversity.

WILDLIFE MONITORING

Wildlife monitoring sometimes involves nothing more than keen observation with no technical equipment at all, as Meriwether Lewis demonstrated. Backcountry hikers, hunters, fishermen, and bird-watchers all notice trends in nature and often report them to park rangers and other professional naturalists. The list below illustrates examples of the trends in nature that have been discovered without advanced technology.

- increases or declines in game animals
- large or small numbers of waterfowl and raptors during migration
- trends in songs from songbirds and hooting by owls
- size and frequency of schools of fish
- diseased trees or discoloration in leaves
- variety of tide pool species
- coral bleaching
- frequency of whale sightings
- limited beach and dune areas

The ivory-billed woodpecker was considered extinct by the early 1970s until two observant naturalists sighted what they believed was the woodpecker in Arkansas in 2003. Ornithologists have since debated the accuracy of recent sightings, but the excitement over the prospect of finding a species presumed to be extinct was undeniable. This event showed that ordinary people could be as valuable in detecting rare species in the wild as trained naturalists. *(Digital sketch by George M. Sutton, Cornell University)*

The chance sighting of an ivory-billed woodpecker (see chapter 1) happened because two kayakers decided one day to explore a secluded swamp. Wildlife protection may never be successful until ordinary people take the time to see nature truly.

People enjoying nature might do well to observe other humans in addition to appreciating plants and wildlife. Ordinary people have the best chance of stopping trouble if they catch someone harassing wildlife. Nature enthusiasts also can be an important deterrent to illegal wildlife hunting, capture, trapping, or interference with wildlife's normal behavior. Other, more advanced techniques of stopping wildlife crime are described in the "How Forensic Science Fights Poaching" sidebar.

HOW FORENSIC SCIENCE FIGHTS POACHING

cologists believe poaching and the illegal trade in endangered species is the world's largest international organized crime after drugs and weapons, and protected areas provide poachers with their best opportunities. Every day park rangers and guards carry out the difficult job of patrolling large reserves in the hope of catching a poacher in the act. Duane Martin, a law-enforcement specialist in Canada's national parks, said in 1992, "National parks are a supermarket of major trophy heads. We grow them protected and we grow them big. We've become a target for these people." Wildlife forensics has developed since the time Martin spoke those words. Park rangers now arm themselves with the same technologies as used by crime scene investigators (CSI) of human homicides.

Wildlife forensics scientists use deoxyribonucleic acid (DNA) typing to solve poaching crimes. DNA typing is based on the genetic makeup of species and individual animals. The DNA of every species on Earth contains small segments that are unique from individual to individual and other segments unique to species. DNA typing matches single strands of these DNA segments—DNA is normally a double-stranded molecule—between two samples. A match between complementary strands indicates that the DNA in both samples came from the same animal, or at least from the same species.

Unfortunately, unlike human homicide scenes, wildlife forensic scientists often come upon a disturbing scene in which very little of an animal has been left by the poachers. They use DNA typing first to identify the species, then perhaps identify the individual animal. Wildlife forensic scientists take samples from the following items to link a specific carcass to a poacher or an animal product: blood in poachers' clothing, bushmeat, pelts, hair, fur, fur coats, reptile leather products (shoes, belts, purses, and wallets), feathers and down, carved ivory products, sea turtle oil (in suntan oil), shell jewelry, and powdered rhinoceros horn.

University-based molecular biologists have in recent years accumulated large databases on the genetic makeup of endangered species by analyzing the DNA of zoo animals. Wildlife forensic scientists depend on this growing store of information to track the route of animal parts

(continues)

(continued)

on the black market. For example, forensic scientists can now trace the route of ivory products back to specific African elephant herds in the wild. Unfortunately, they have a hard time catching up with poachers, who often hunt at night and use their own tricks to find their prey and evade the law.

Wildlife forensic science has kept up with the latest advances used by CSIs and at the Federal Bureau of Investigation. In addition to DNA typing on nuclear DNA (found in cell nuclei), they also use mitochondrial DNA analysis, polymerase chain reaction, and other sensitive tests to solve wildlife crime.

Wildlife detective Jim Banks has helped catch illegal hunters of elk, deer, and other game animals in California with a mere snip of sample from a killed animal. Game Warden John Dawson has told suspected poachers, "I tell them we have the best forensic scientist in the nation who will be able to tell me the exact sex and the number of animals represented by a blood smear or meat sample." Many poachers are shocked by this news and confess on the spot.

THE DELISTING PROCESS

An animal species placed on the endangered species list follows one of two paths: a continued downward slide to extinction or recovery of the population so that the U.S. Fish and Wildlife Service (FWS) delists it. Delisting is the removal of a species from the endangered species list because its population has been restored to a sustainable level and is no longer endangered or threatened (or because it is known to have gone extinct despite protections). The goal of the Endangered Species Act is to recover an animal or plant population sufficiently to delist it. *Downlisting* is moving a species from one category to a less threatened category. For example, when an animal moves from endangered status to threatened status, it has been downlisted.

Three things must happen before a species can be considered for delisting. First, FWS officials determine that the species' populations are no longer declining. Second, the population must stabilize—it can maintain its numbers without human help. Third, the species must increase in number to prove to officials that it can survive on its own in the wild.

After the three steps described here have been taken, the FWS does a five-point evaluation to decide if the species still needs protection, as follows:

1. Is the species' habitat or range presently destroyed or threatened?

2. Is the species likely to be sought after for commercial, educational, or other purposes?

3. Is the species threatened by disease or predation?

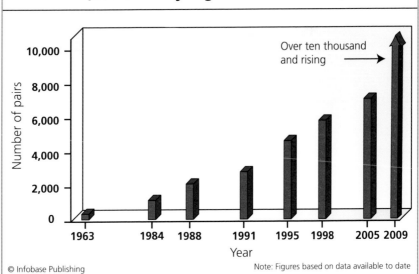

Bald Eagle Nesting Pairs (Lower Forty-eight United States)

© Infobase Publishing

Note: Figures based on data available to date

Species protection laws and the U.S. ban on DDT have enabled nesting pairs of bald eagles to increase in the continental United States. The U.S. Fish and Wildlife Service delisted the bald eagle from the endangered species list on June 28, 2007. Protections continue under the Eagle Act, which controls killing or injuring of eagles or their nests. Eagle numbers continue to increase.

4. Are there inadequate legal protections outside the Endangered Species Act?

5. Are other natural or human factors affecting the species' existence?

If the FWS answers no to all of these questions, it then makes public its intent to delist the species. The public, scientists, and government offices have several weeks to comment on the delisting plan. After all the comments have been collected, the FWS publishes its decision to delist the species, and the agency removes the name from the list. Downlisting follows the same steps. The FWS monitors all delisted animals for at least five years to assure that no changes in habitat or health have occurred to affect its survival.

Almost 50 species of animals and plants have been delisted from the endangered species list since its inception. The following tables show the final outcome of these species.

Animals that have been delisted because their populations recovered are the following (recovery periods in parentheses): Yellowstone grizzly bear, bald eagle (40 years); American alligator (8 years); peregrine falcon (29 years); Aleutian Canada goose (34 years); brown pelican, Atlantic coast (15 years); gray whale, Pacific coast (14 years), and gray wolf (Rocky Mountain, Minnesota, and Great Lakes populations). President George W. Bush removed the gray wolves of the upper Midwest from the endangered species list in early 2008, but the FWS sought

ENDANGERED SPECIES ACT DELISTING RESULTS (1970–2010)			
CATEGORY	NUMBER RECOVERED	NUMBER EXTINCT	REMOVED FOR OTHER REASONS*
Plants	3	0	9
Animals	17	9	9
* Usually this is because of a change in the taxonomic classification of the species.			

ANIMALS DELISTED FROM THE ENDANGERED SPECIES LIST BECAUSE OF EXTINCTION		
ANIMAL	DATE LISTED	DATE DELISTED BECAUSE OF PRESUMED EXTINCTION
Amistad gambusia (fish)	April 30, 1980	December 4, 1987
blue pike (fish)	March 11, 1967	September 2, 1983
dusky seaside sparrow (bird)	March 11, 1967	December 12, 1990
Guam broadbill (bird)	August 27, 1984	February 23, 2004
longjaw cisco (fish)	March 11, 1967	September 2, 1983
Mariana mallard (bird)	December 8, 1977	February 23, 2004
Sampson's pearlymussel (crustacean)	June 14, 1976	January 9, 1984
Santa Barbara song sparrow (bird)	June 4, 1973	October 12, 1983
Tecopa pupfish (fish)	October 13, 1970	January 15, 1982

to return it to protected status later that year. See the case study "Yellowstone's Wolves" in Chapter 3 for a discussion on the controversies that surround wolf protections.

CONCLUSION

Any species vulnerable to extinction needs protections so that it does not become unrecoverable by conservation biologists. Species protections therefore are a set of activities that may not be natural to a species—defined wildlife reserves, zoos, aquariums, sanctuaries, and the lake—but these approaches provide the only means of saving biodiversity that has already been severely damaged by humans.

Species protection includes a combination of wide-reaching international laws and local efforts. Either internationally or locally, protection efforts now incorporate many disciplines in order to succeed. Preserving endangered species has become a matter of economics, social and cultural values, politics, and even religion in addition to biology. This is because there are few places on the Earth not affected in some way by human activities and human behavior. Biodiversity suffers today from extremes in poverty and affluence. These conditions directly or indirectly cause species loss.

International organizations have helped get various nations to cooperate on halting the worst activities that cause biodiversity loss, but much more cooperation will be needed to divert species from extinction. Large, well-funded organizations can use sophisticated methods for monitoring the status of the world's species, and ordinary citizens can join local efforts to protect wildlife, but wildlife protections remain hampered by a combination of cultural values, myth, misinformation, and real economic needs. Even the delisting process for taking an animal off the U.S. endangered species list creates heated arguments from time to time when environmental concerns oppose economic concerns.

The technologies for saving biodiversity have improved over the past several decades, but technology may not be the most difficult part of biodiversity programs. The difficult aspect of saving biodiversity and protecting species comes from the need to create open communications among local and national governments, environmentalists, scientists, large and small businesses, and even religions. In other words, human cultural values impact species protections as profoundly as they affect wealth, poverty, war, and peace.

METHODS FOR MEASURING DIVERSITY

Biologists measure biodiversity and its loss using the methods discussed throughout this book. Some measurements offer very precise data and other methods produce estimates. Almost all biodiversity studies depend on mapping techniques. Three of the most useful maps used by biologists are those that show an area's species richness, rarity, and endemism. Richness is the number and frequency of species in a given region. Rarity is a measure of the infrequency of species. Richness and rarity define biodiversity perhaps better than any other measurements in biology. Finally, endemism refers to a situation in which a species is found nowhere else but in one confined region. Endemic species signal the status of biodiversity in a region, because if their numbers decline quickly, these animals will be irreplaceable.

Many animal species remain difficult to monitor. For example, most reptiles live out of sight in inhospitable places, but counts of big reptiles such as crocodilians, turtles, and tortoises are easier. Also, some bird and mammal species migrate and so make the final tally of a species within an area difficult to determine.

This chapter reviews the types of measurements biologists take to define richness, rarity, endemism, and other characteristics of habitat. It covers not only measurements for terrestrial animals, but also plants and marine organisms.

CLIMATE AND TOPOGRAPHY

Ecology includes the mapping of climate and land topography around the world. These two physical features help identify the biomes. The features of biomes create a clearer picture of large, habitats and also smaller, specialized habitats called *microenvironments*. For instance, soil is a habitat for many invertebrates and microorganisms, but the oxygen-free soils located under bogs and swamps make up microenvironments. Mapping and monitoring therefore cover a wide range of territories. Ecological studies begin with global scans of biomes and then focus on smaller, more specialized places until they reach single microenvironments.

The Earth's biosphere contains the parts of the planet where life exists (almost everywhere). The biosphere contains the atmosphere (troposphere), surface and underground waters (hydrosphere), and the soils and sediments of the Earth's crust (lithosphere). Biology and ecology focus on areas within the biosphere that define life and the many roles played by living things, particularly the energy-matter cycle described Chapter 1.

In environmental science, the first step involves an understanding of what will be measured and what these measurements will reveal about the environment. Measurements taken to define biomes produce different final results than a single temperature or pH measurement, for instance, of a microenvironment. Ecological studies often start from the large scale and progress to the small, in other words, from biomes to unique ecosystems. Biomes contain living and nonliving components in a terrestrial area; deserts, grasslands, and forests are examples of biomes. *Biogeographic regions* are similar to biomes, but they contain more detail regarding climate and topography, which in turn determines the types of vegetation growing in them. Vegetation determines a range of animal species most likely to thrive in the area. Each biogeographic region, such as a large island or a continent, contains communities. A community is a collection of all the species living and interacting in an area. Individuals of the same species form organizations within each community known as populations. Populations may contain smaller groups of individuals such as packs, herds, or flocks.

Communities in a specific shared region may also make up an ecosystem. A freshwater lake provides a simplified example. In a lake, microbial

(opposite) Ecologists have classified all the regions on the planet according to climate and topography, which influence the types of plants and animals living in these biogeographic regions. By studying the characteristics and environmental condition of each region, ecologists can draw conclusions on the threats to species and habitats.

Biogeographic Regions

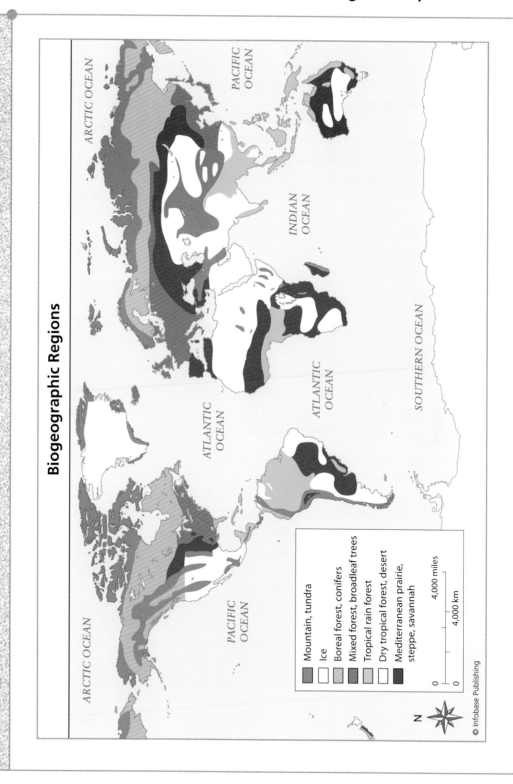

Mountain, tundra

Ice

Boreal forest, conifers

Mixed forest, broadleaf trees

Tropical rain forest

Dry tropical forest, desert

Mediterranean prairie, steppe, savannah

0   4,000 miles

0   4,000 km

ARCTIC OCEAN

PACIFIC OCEAN

INDIAN OCEAN

ATLANTIC OCEAN

ATLANTIC OCEAN

SOUTHERN OCEAN

ARCTIC OCEAN

PACIFIC OCEAN

N

and invertebrate communities swim in the water, underwater plant communities grow along the submerged surfaces, and plant and insect communities gravitate toward the water-air interface at the water's surface. The entire ecosystem contains these communities in addition to populations of fish, amphibians, and waterfowl.

Climate type and topography both define the features of biogeographic regions. Climate is a region's general weather patterns that occur over an extended period of time, usually a year. In 1928 Russian-German climatologist Wladimir Köppen devised a climate classification system that closely mirrors the biogeographic regions ecologists use today. The Köppen Climate Classification System contains the following categories:

- tropical moist—all months have average temperatures above 64°F (18°C); includes local climates called tropical wet, tropical monsoon, and tropical dry (savanna)

- dry—deficient precipitation most of the year; includes local climates called dry arid (desert) and dry semiarid (steppe)

- moist midlatitude with mild winters—warm, humid summers and mild winters; includes local climates called humid subtropical, Mediterranean, and marine

- moist midlatitude with cold winters—warm to cool summers and winters less than 37°F (3°C); includes local climates called dry winters, dry summers, and all-season wet

- polar—year-round cold with warmest temperature less than 50°F (10°C); includes local climates called polar tundra and the polar ice caps

Two factors influence climate: ocean currents and topography. Ocean currents and the winds they create exchange heat with the atmosphere and also transfer heat across the planet's surface. Land masses exchange heat more slowly than water, partially due to the land's topography. Topography consists of the physical and natural features of the land's surface; it influences climate because land at different elevations holds and releases heat differently. Topography also affects the type of soil, moisture levels, humidity, and sunlight penetration that ecosystems receive.

MAPPING SPECIES AND GROWTH PATTERNS

Geographers, climatologists, and ecologists handle large databases that have accumulated years of information on weather, precipitation, biota, human populations, surface waters and flow, and soil types. Several years ago scientists collected data on all these things by hand tabulations. Today they use global positioning systems (GPS) to determine the exact position of the small areas being measured. Geographic information systems (GIS) have become even more useful than GPS. GPS tells users their exact geographic location, but GIS includes other pieces of information in large databases. GIS calculates and analyzes data so that scientists can create maps of complex things such as biodiversity, natural resources, different types of pollution, and global sea temperatures. In 2007 the Intergovernmental Panel on Climate Change (IPCC) released a report on the latest effects of global warming from data generated from global mapping. "Warming of the climate system is unequivocal," the report said, "as is now evident from observations of increases in global average air and ocean temperatures, widespread melting of snow and ice, and rising global mean sea level." Ocean depth, currents, and temperature maps have now been applied to concentrated studies in select parts of the world. For instance, Gilly Llewellyn of the World Wildlife Fund described to the *Australian* newspaper in 2006 studies being conducted on the Southern Ocean and Antarctic marine habitats: "This map will help us to better understand the Southern Ocean so that we can address the major threats to its habitat and wildlife, such as illegal fishing, climate change and the impact of invasive marine species." Environmental science continues to blend simple manual methods of study with advanced electronic technologies.

Mapping methods target two features of individual animal species: range and incidence. Range is the widest area where a population of animals roams, and incidence is the number of times animals occur in their range. Biologists create separate range maps and incidence maps, then put one atop the other for each species. In this way the maps show the density of a species throughout its range. These range-incidence maps can be made to cover areas as large as 250 square miles. The large-area maps consist of many maps of small areas, called grids, compiled into a single overview of a large region. For example, the range-incidence map of California contains 300 individual grids.

Richness measurements can also be mapped to give biologists a close approximation of an area's biodiversity. Scientists in fact often refer to richness as a measurement of diversity. Rarity, by contrast, indicates any

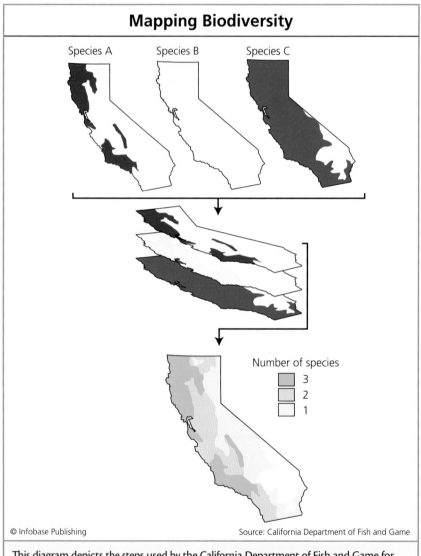

Mapping Biodiversity

© Infobase Publishing

Source: California Department of Fish and Game

This diagram depicts the steps used by the California Department of Fish and Game for measuring and mapping biodiversity. Biologists first divide the state into 300 equal area regions, then teams count all the species they find in each region. The biologists then construct maps to show the range where each species is found. By overlaying all the maps, they build a biodiversity map to indicate high, moderate, and low regions of biodiversity.

threats to biodiversity in an area. Areas with the highest rarity values are those in which all species exist at a very low incidence. Environmental organizations such as the IUCN construct maps of today's evolving hotspots by focusing on richness, rarity, and endemism. The "Biodiversity Hotspots around the World" sidebar discusses these places of amazing biodiversity.

PLANT DIVERSITY RICHNESS

Diversity in single-celled and multicellular plants supports almost all food chains. Plant diversity influences the animal populations living in an area, as described in the sidebar "Species Succession" on page 167. Unfortunately, human activities have overburdened the world's producers (plants), and this has contributed to animal extinction. Humans use 3 percent of the energy stored in plants for food, feeding livestock, and heating by the equation shown below. When people convert vegetated land to nonbiological uses—parking lots, buildings, marshes to dry land, grasslands to deserts—this amount increases to 30 percent.

$$\text{Solar energy} + \text{carbon dioxide} + \text{water} \rightarrow \text{glucose} + \text{oxygen}$$

The total amount of the Sun's energy held by plants is called gross primary productivity (GPP). The rest of the Sun's energy dissipates as heat and is lost. Producers hold 100 percent of the energy they capture in their tissues. They use some of it for building plant tissue, also called maintenance-growth-reproduction (MGR), and the rest goes to provide energy for every other living thing. This remainder on which animals depend is called net primary productivity (NPP).

$$NPP = GPP - MGR$$

Plant diversity runs the planet because different plants serve as producers in hundreds of food chains. Biologists measure plant diversity richness by analyzing the types of plants known to live in habitats within each biogeographic region. Plant range-incidence mapping shows that the United States contains plant diversity hotspots in the San Francisco Bay region, Hawaii, the Mojave Desert, the Smokey Mountain region of the Appalachian Range, and spots in Southern California.

BIODIVERSITY HOTSPOTS AROUND THE WORLD

British biologist Norman Myers first introduced the idea that biological diversity is linked with extinction. In 1988 Myers described 10 select areas around the world that contained very high levels of diversity and called these places biodiversity hotspots. Biodiversity hotspots contain the greatest variety of endemic species on Earth, but they also undergo rapid habitat loss. "The grandscale of species loss may not be the full story," Mr. Myers has pointed out. "The current biotic crisis is grossly depleting the capacity of evolution to generate replacement species within a period of less than five million years." The ecological costs of this extinction rate may last, according to Myers, "for a period of at least twenty times longer than humans have been humans." Myers and his contemporaries realized that hotspots represented the places most in need of conservation in order halt premature extinction.

Conservation International provides information on all the world's hotspots and has set criteria for defining any region as a hotspot. The two main criteria relate to habitat quality: a region must contain 1,500 different endemic vascular plants, and the region must contain 30 percent or less of its original vegetation.

Thirty-four hotspots have now been identified around the globe (see Appendix C). The following table shows the distribution of hotspots across the world's biomes. (The total is more than 34 because some biomes contain more than one hotspot.)

The 34 hotspots once covered 15.7 percent of the Earth's surface. They now cover 2.3 percent, but 77 percent of the world's species live in them. Even more troubling is the problem of war; many hotspots lie in regions of violent conflicts or all-out war. Even without war, population encroachment poses an inexorable threat. Hotspots in Japan and the Philippines, for example, contain 874 and 710 persons per square mile, respectively (336 and 273 persons/km^2). Added to all those factors, climate change affects even remote hotspots such as southwestern Australia, which holds no more than 13 persons per square mile (5 persons/km^2).

PLANT RARITY

Rarity is the opposite of richness; it is a measure of species infrequency. Scientists calculate rarity by taking the inverse of richness, a conversion called a *rarity-weighted richness index* (RWRI). To calculate this index, each plant species found in a study area receives a numerical value. This value comes from the inverse of the number of areas in which the plant occurs by completing the following steps:

Even with added attention to biodiversity hotspots, species continue to disappear at a rate that may be several thousand times greater than the natural rate of extinction. Michael Hoffman of Conservation International said in *National Geographic* in 2005, "The hotspots strategy really intends to pick out regions across the globe where we need to go first to be effective in saving species." The hotspot approach, therefore, will not automatically save biodiversity, but it will help the places that have the most to offer Earth's vast richness of plant and animal diversity.

THE WORLD'S HOTSPOTS ACROSS BIOMES	
BIOME	**NUMBER OF HOTSPOTS**
tropical, subtropical moist broadleaf forest	17
montane grasslands and shrublands	5
Mediterranean forests, woodlands, and shrubs	5
temperate broadleaf and mixed forests	5
tropical, subtropical grasslands, savannas, and shrublands	2
tropical, subtropical coniferous forests	2
temperate coniferous forests	2
deserts and xeric (requiring little moisture) shrublands	1
tropical, subtropical dry broadleaf forests	1

1. Record the number of grids where a species is observed by using a range-incidence map.

2. Divide 1 by the number of grids where the species was observed. (A species inhabiting only one grid receives a score of 1.0; a species inhabiting 20 grids receives a score of 0.05.)

3. Calculate RWRI for each grid by summing the scores of each species in the grid. As the RWRI number increases, so does a

plant species' rarity. Therefore, the plant rarity value may be thought of as a measure of irreplaceability, because the plants scoring the highest are considered the rarest. Conservation biology currently focuses on plants with high RWRI scores.

The world contains plant rarity hotspots the same way it houses plant biodiversity hotspots. Surprisingly, rarity occurs not in places where biodiversity is poor, but in places with high diversity. The largest number of rare plants on Earth occurs in places with rich biodiversity and also in places of large human population size.

The United States contains about 8,840 rare plants that make up one-third of the country's total flora. Some plants are naturally rare and may not be in danger of extinction. Others become rare due to fragmented habitat, loss of compatible soils, seeds that do not disperse well in the environment, or seeds that depend on certain insects or pollinators that are themselves endangered. The ecosystem approach to biodiversity loss focuses on the entire plant-animal population in any given area. In this approach, the plant richness and rarity composition plays a critical role.

ANIMAL DIVERSITY RICHNESS

Richness maps of mammals, birds, amphibians, and reptiles help in assessing the overall health of the environment. Since richness represents the number of species found in an area, it tends to be high in places where urbanization is low. Biologists compile animal richness data from field studies using manual counts of species in the wild. Over time, many studies by hundreds of biologists produce a large store of information on the world's species that go into making accurate maps.

Animal richness calculations can be done for entire continents, states, or even one's backyard. A simple species richness calculation using some of Michigan's native animals as examples is shown in the following table. In this example, areas 2 and 4 have little species richness, but areas 3 and 5 contain rich animal diversity.

ANIMAL RARITY

Rare animals, like rare plants, are not necessarily endangered, because every animal is rare somewhere in the world. Migration complicates ani-

SPECIES SUCCESSION

cological succession is a process of new organisms colonizing an area in which increasingly complex organisms succeed each previous set of organisms. Succession takes place when plants recolonize an area damaged by fire, for instance, or when ecologists restore ecosystems and habitat. Plants colonize the Earth in a rather uniform manner that is well understood in biology, but animal succession may take place in more than one way.

The type of vegetation growing in a place at any particular time helps biologists predict the types of animals likely to inhabit the same area. In addition, different animal species dominate an area during each phase of plant succession. One theory that explains how animal succession works in an evolving habitat is called the *habitat accommodation model.* This theory proposes that animals enter an area undergoing succession only when the vegetation there can support them, and as plant succession progresses, these animals leave the area and others that use the vegetation replace them.

Studies on animal variety and abundance in areas undergoing plant succession show the following patterns in animal succession:

1. At 0–5 years when pioneer mosses and lichens begin growing, insects enter, followed by insect-eating birds and small rodents.

2. At 6–25 years during the shrub and bush phases, birds, reptiles, rodents, and other small mammals use the dense vegetation for cover and nesting. These attract predators such as foxes, coyotes, weasels, and bobcats. Taller shrubs and saplings draw deer. Large predators such as mountain lions and bears then move in.

3. At 26–50 years a young forest of deciduous trees provides a forest canopy, and shrubs and plants decrease at the forest floor. Most of the animals up to this point move out, but beavers use fallen trees and certain bird species use the forest canopy.

4. At 51–150 years evergreens dominate the mature forest, and deciduous trees begin to die and fall, leaving openings in the canopy. Shrubs and short plants take advantage of the sunlight, and birds and small mammals return.

5. At 151–300 years at climax forests of large evergreens grow evenly spaced from each other. Dead trees provide nesting for woodpeckers, owls, and other animals that live in or under fallen trees and logs. Depending on the activity of decomposers and the amount of deciduous trees, additional animals may use the habitat. Nesting birds such as warblers move in, along with a small number of predators such as owls and hawks. The numbers of most small and large mammals stay low, but chipmunks, porcupines, and deer dominate. Reptiles are rare, but amphibians such as salamanders dominate the undergrowth.

Global Biodiversity Hotspots

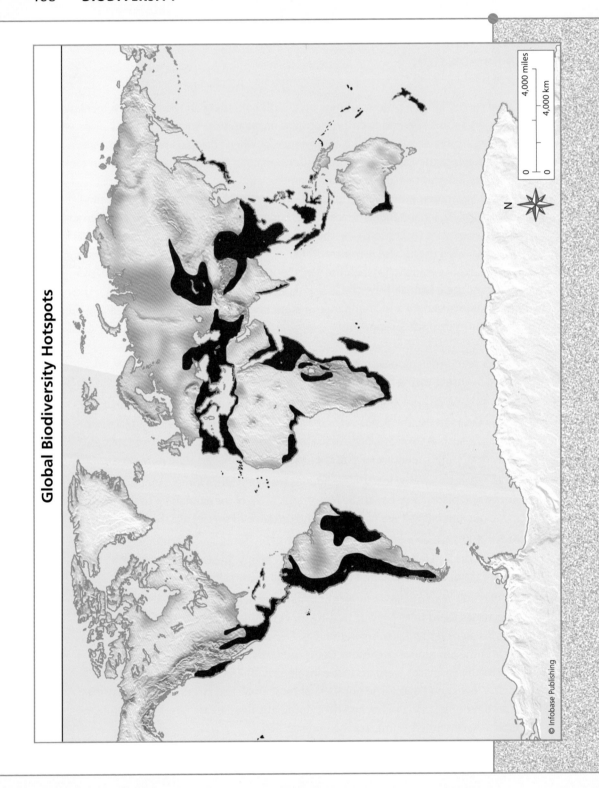

© Infobase Publishing

	INDIVIDUAL ANIMALS COUNTED				
SPECIES	**AREA 1**	**AREA 2**	**AREA 3**	**AREA 4**	**AREA 5**
gray wolf	0	2	0	0	0
fox	1	1	2	0	0
black bear	0	0	2	0	0
deer	7	3	6	0	12
bat	2	0	0	0	5
heron	0	0	0	0	5
goldfinch	0	0	9	2	2
totals	10	6	19	2	24

HOW TO CALCULATE SPECIES RICHNESS

mal rarity because there may be times when a wetland is almost covered bill-to-tail with snow geese, yet almost on cue, thousands of geese rise into the air to begin their migration. Within a week the wetland that held uncountable numbers of geese now contains one or two, or none.

In addition to natural migrations, some animals roam over expansive ranges, and so rarity calculations must account for this movement. Consider a species (species A) that occurs in four of five study areas; it therefore has a large range. A second species (species B) may have been spotted in only two of five areas; it has a smaller range. Range-size rarity is the inverse of the number of areas where a species has been

(opposite) This map shows the locations of the world's biodiversity hotspots, which generally increase toward the equator. Most of Earth's regions that hold the greatest diversity of plant and animal species are also under the greatest threat from population encroachment, pollution, and other types of habitat destruction.

counted. For the two species in this example, range-size rarity values are as follows:

Species A range-size rarity = 1/4 = 0.25
Species B range-size rarity = 1/2 = 0.5

For individual species, a higher index number signifies greater rarity. The range-size rarity for the single-study area above equals 0.75. Range-size rarities for more than one hypothetical study area are illustrated in the two following tables.

In calculating an area's range-size rarity, rather than single species, higher values indicate areas with greater diversity (less rarity). The range-size rarity calculation therefore accounts for the range of an animal in addition to its abundance.

Animal migrations and movements over a range contribute to an animal's RWRI. Rarity-weighted richness values become most useful when compared over a period of time to determine whether a species is growing more abundant or slipping toward extinction. It should not be difficult to imagine the rarity indexes (if they had been calculated) of American bison on the Great Plains in 1775, 1830, and 1995. In other words, as the years progressed, the bison's RWRI increased.

AMPHIBIANS AND REPTILES

Biologists calculate richness and rarity in amphibians and reptiles as they do for plants and other animal life. Finding these species to count them

CALCULATION OF RANGE-SIZE RARITY FOR AN AREA					
	NUMBER OF SIGHTINGS OF EACH SPECIES IN EACH AREA				
SPECIES	AREA 1	AREA 2	AREA 3	AREA 4	AREA 5
A	1	0	1	1	1
B	0	1	0	1	0

SPECIES RANGE-RARITY AND AREA RANGE-RARITY					
	SINGLE SPECIES RANGE-RARITY VALUES				
SPECIES	AREA 1	AREA 2	AREA 3	AREA 4	AREA 5
A	0.25	0	0.25	0.25	0.25
B	0	0.5	0	0.5	0
area range-size rarity totals	0.25	0.5	0.25	1.50	0.25

presents a challenge, however, because they have evolved superb methods of going undetectable to predators: camouflage; ability to stay completely still; ability to hide underwater, in soils, and in mud; or the preference for staying under rocks and tree trunks or in trees. Ecology research uses two monitoring methods for amphibians and reptiles—visual counting and capture—but it also requires scientists to have a good sense of where to find these animals.

Two common visual methods for monitoring amphibians are migration counts and sit counts. Migration counts involve monitoring a location at night during migration season and recording the number of animals that pass through a counting area, for instance, animals crossing a road. The second method—equally as tedious—involves sitting in a single location and recording each species seen in a set time period.

Some species have characteristics that make counting them, if not easier, more fun. One example is the frog call survey. In this method a biologist drives along a route in the evening and stops the engine every half mile to listen for five minutes to frog calls. Volunteers become fairly good at telling how many frogs or toads they hear calling in the woods. Geckos provide a second specialized monitoring method. Some geckos fluoresce in ultraviolet light, so biologists take advantage of this feature when tramping through the woods with an ultraviolet lamp in hand.

Capturing amphibians and reptiles makes monitoring easier than stalking them in the night. Commonly used capture methods are the following: hand-capture, trapping, artificial refuges, and marking-recapture.

Hand-capture is useful in reptile surveys involving snakes, lizards, tortoises, and aquatic turtles and terrapins. Biologists may capture reptiles hiding under logs, rocks, scrap metals, and derelict cars, or they can use mirrors attached to long sticks for checking into cracks and holes. Reptile capture requires a modest amount of training to avoid bites and to handle properly species that exhibit tail autonomy, meaning they can drop their tail if caught. It helps also to remember a cautionary rhyme when catching snakes, "Red and yellow kill a fellow. Red and black, venom lack."

Biologists capture crocodiles, alligators, and large lizards by noosing. Noosing involves a long, stout stick with a noose at one end. Once the animal is snared, the noose can be operated to create a tight hold on the animal, which can be counted on to put up a fight. Biologists use dip nets rather than nooses for capturing aquatic species such as terrapins and turtles.

Two types of traps can collect amphibians and reptiles without the biologist being present. The first, a pitfall trap, is simply a bucket dug into the ground so its mouth is even with the surface and covered with camouflage twigs, sticks, or leaves. Funnel traps are the second method in which an aboveground funnel-shaped container collects the animals. Short-drift fences aid both methods by providing a barrier that guides the animals toward the trap. Trapping does not harm the animal and scientists release the trapped individuals after the counting period ends.

Artificial refuges also collect animals, but they do not usually trap them. Any open-ended canopy made of a sheet of wood or metal provides a shady, dark habitat that attracts amphibians and reptiles. Well-constructed artificial habitats allow scientists to count and study the animals but not affect their behavior.

The marking-recapture method uses special, nonpainful techniques for marking individual animals so that they are recognized if recaptured. Common marking methods are toe-clipping (for species that do not climb trees), scale removal, painting, or passive integrated transponder (PIT) tags placed under the skin and monitored with a handheld scanner. Statistical calculations result in data containing the following two components: the number of animals in an area, based on the number of animals captured, and the number of times the same animal has been captured.

AQUATIC BIODIVERSITY

Aquatic species richness and rarity measurements depend on techniques similar to those used on land. Fish are mobile animals that go through several life-cycle stages. Monitoring methods must take these factors into consideration; usually fish counts are done on a specific age. Small pools and clear streams lend themselves to visual surveys, temporary catches, or traps. Migrating fish in larger, fast-flowing streams may require artificial structures that funnel them through a specific area. Biologists then visually count them or use an electronic device.

In deep, still waters, marine biologists monitor species using sonar readings or a variety of netting techniques, especially gill nets, trawl nets, and seine nets. Gill nets consist of mesh that allows in fish of only a certain size. Smaller fish swim back out of gill nets and larger fish do not fit into the mesh openings. Collections may also be done with trawl nets that a boat (trawler) drags by a long cable through the water or along the water's bottom. Seine fishing nets have weights that hold the net's middle area on the water's bottom and allow the open mouth to reach the surface. Fishermen bring in their catch by drawing on ropes until the net closes, like closing a cloth sac. Trawling and seine fishing are not usually selective; they catch fish of various sizes.

Aquatic invertebrate monitoring requires specialized techniques because these species contain a wide range of size, life cycles, and microenvironments. Some of the most common monitoring methods are the following: aquatic vegetation sampling to count the species attached to aquatic plants; net sampling for species suspended in the open water; kick sampling to dislodge species living among the gravel, sand, and pebbles at the water's bottom; cylinder sampling to collect a set volume of water and bottom matter; and artificial substrates onto which invertebrates attach and grow over time.

Aquatic algae and other simple organisms usually require microbiology methods. One collects and dilutes a water sample, then applies it to a mixture containing nutrients. After an incubation period, a microbiologist counts the number of cells per unit of sample volume.

Marine biologists often resort to diving or snorkeling to capture representative aquatic samples. In general, aquatic biodiversity monitoring uses specialties that require extra training. The case study "Ocean Warming and Phytoplankton" describes technologies adopted for global studies of aquatic conditions. Though all biodiversity measurement depends on estimates rather than actual counts, aquatic methods present the most difficult challenges of all monitoring methods.

CASE STUDY: OCEAN WARMING AND PHYTOPLANKTON

Plankton is the name for free-floating cells that live in open waters. Zooplankton consists of animal cells and phytoplankton consists of plant cells. Phytoplankton performs photosynthesis and may be the world's most important producer in the earth's energy-matter cycle; it is the foundation of the majority of ecosystem food chains.

Phytoplankton levels in the ocean increase when water temperatures are cool and decrease when temperatures warm. Ocean temperatures have followed a warming trend over the past 50 years, and so phytoplankton are declining and aquatic ecosystems struggle with this decrease in aquatic biodiversity. Phytoplankton in the ocean plays another key role in the environment by consuming carbon dioxide from the atmosphere. Therefore the loss of phytoplankton is a two-fold problem: accelerated buildup of greenhouse gases and biodiversity loss.

Unlike the measurements used for monitoring other forms of life, scientists get their best view of the world's ocean phytoplankton levels through satellite imaging. They determine the amount of phytoplankton in the ocean by measuring its color with specialized satellites. Marine scientist Dave Siegel told the *San Francisco Chronicle* in 2006, "We can know the amount of plants in the ocean by looking at the ocean's color. A blue ocean has no phytoplankton in it. The beautiful tropical oceans that you see on postcards have little in it. The green ocean is chock-full of phytoplankton."

Marine biologists research phytoplankton to answer questions from the following subject areas: characterizing the overall diversity of phytoplankton in the global oceans, understanding aquatic ecosystem structure and function, determining the contribution of marine photosynthe-

RANKING THE RARE SPECIES

After biologists determine the richness or the rarity of species in a habitat, they rank the species according to factors related to vulnerability. Put another way, various species fall into different categories based on the degree of risk of going extinct. Such rankings lead to lists such as the endangered species list and the Red List.

The Red List contains more than 40,000 species, estimated to be about 12 percent of the world's total species. The list places today's existing species into any of five categories based on key criteria: population size, population growth or decline, habitat fragmentation, and reproductive rate.

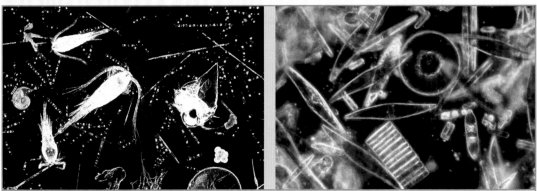

Phytoplankton consists of a highly diverse group of tiny invertebrates and algae that form the foundation of aquatic ecosystems. Programs that protect marine water quality also help maintain healthy phytoplankton populations. (a) the genus *Cerataulina* *(OAR/National Undersea Research Program)* and (b) marine diatoms *(Neil Sullivan, NOAA/Department of Commerce)*

sis to ecology, learning about interactions of microscopic organisms in ecosystems and marine communities, and conducting further studies on the effects of global warming on ocean ecology. In a big-picture sense, understanding phytoplankton enriches science's understanding of the life that evolved in the oceans in Earth's early history.

Species protections begin with tools such as the Red List and the endangered species list to identify species facing the most severe threat of extinction. Polar bears, for example, are expected to decline by 30 percent in the next 45 years according to the United Nations IUCN. *(Jurgen Holfort, International Arctic Research Center, University of Alaska-Fairbanks)*

(Though species is used here, on occasion the Red List classifies animals or plants in their genus or family; see the sidebar "Carl Linnaeus—The Father of Taxonomy" in chapter 1.) The listing also takes into account how much land an individual animal needs to sustain an entire population of its species. For instance, mountain lions require areas measured in square miles, but some songbirds live their entire lives within the same acre.

The following table shows only the Red List's three most at-risk classifications: critically endangered, endangered, and vulnerable. The list also includes near threatened species and species of least concern. Animals classified as near threatened are not in immediate risk, but science holds evidence that they are close to being vulnerable, endangered, or critically endangered. Widespread species in abundance in their habitat belong to a category called least concern. (Other terms used informally are critically imperiled, imperiled, vulnerable, secure, and widespread to correspond with the Red List's terminology.) With only minor differences, the U.S. Endangered Species Act listing and the Canadian Wildlife Service's Species At-Risk Act match the criteria used in the Red List.

The Red List provides another piece of information not included on the U.S. endangered species list: all available data on extinct species. The list contains two categories of extinct species: (1) extinct means there is no reasonable doubt that the species is gone, and (2) extinct in the wild means a species is believed to exist only in captivity or in cultivation (plants).

No list can solve the problem of biodiversity loss, but endangered species lists clarify the problem for scientists and for the public. Jane Smart, the head of the IUCN's species program, said to the United Kingdom's *Guardian* newspaper in 2007, "Our lives are inextricably linked with biodiversity and ultimately its protection is essential for our very survival." Ecology is truly a science that discovers the interrelationships among living things.

	IUCN RED LIST CRITERIA FOR THE WORLD'S SPECIES		
CRITERIA	**CATEGORIES**		
	CRITICALLY ENDANGERED	**ENDANGERED**	**VULNERABLE**
reduced population size	≥ 90%	≥ 70%	≥ 50%
small range	< 100 km^2	< 5,000 km^2	< 20,000 km^2
small area used within a range	< 100 km^2	< 500 km^2	< 2,000 km^2
small and declining population	< 250	< 2,500	< 10,000
very small population (mature individuals)	< 50	< 250	< 1,000
very small range for vulnerable species			< 20 km^2 or ≤ 5 locations

SPECIES DISTRIBUTION

Species distribution describes the pattern in which species disperse over a region. Crows, for instance, have a wide distribution across most continents. Conversely, narrow distribution occurs when a species lives only in a small area, such as manatees in the waters of central and south Florida.

Distribution comprises two additional characteristics: evenness and *dominance.* An even distribution results when a similar number of individuals live in different areas. Species that live in social structures (herds, packs, flocks, schools, etc.) tend to travel as groups, and so they do not have evenness, but rather live in a *clumped distribution.* Groups such as these protect their members against predation, aid in hunting, and help in breeding and raising young. The second type of distribution comes from dominance. Dominance occurs when biologists see more individuals of a species than expected. Dominance may be an indication of an imbalanced ecosystem. Dominance may also arise from temporary conditions due to fire, flooding, or a sudden influx of people into a habitat.

Species distribution consists of many aspects that scientists understand and likely some aspects that are not yet understood. A brief list of the factors that influence species distribution are the following, all topics that have been discussed in this book: biome characteristics, climate change, ecosystem health, habitat status, invasive species, competition and adaptation, status of migration corridors, vulnerability of species to illegal killing and trade, urbanization, and special protections for endangered and threatened species.

CONCLUSION

All studies of biodiversity depend on data so that scientists can assess biodiversity hotspots and biodiversity loss. These data accumulate after years of scientific studies on individual species. In measuring biodiversity of plants, animals, insects, aquatic species, or any other living thing, richness and rarity are key pieces of information. Richness is the number of individuals of each species in a given region. Ecologists evaluate richness to assess biodiversity. Rarity, the infrequency of a species, occurs if an animal normally exists in low numbers or is nearing extinction.

Most measurement techniques are labor-intensive methods called field studies that depend on scientists' going into natural settings to catch, trap, or listen to animals. Field studies give biologists estimates of animal numbers, but after many years of accumulating data on species, their estimates can be very accurate for terrestrial animals. The methods for aquatic species are similar, but so far scientists have accumulated less information about these species.

All of the measurements described in this chapter help environmental scientists rank species according to their likelihood of premature extinction. Different organizations use slightly different terminology in their ranking systems, but overall, they agree with five major groups: critically endangered, endangered, threatened, near threatened, and secure. Biodiversity protections will depend on further advances in the technologies scientists use for assessing species. With each advance, environmental science can develop more useful and meaningful protections for individual species, their habitats, and as a result, global biodiversity.

8

FUTURE NEEDS

Understanding biodiversity is a complicated task simply because of the immense amount of data that science has compiled on the world's species and perhaps the far greater amount of information still undiscovered. Biodiversity and climate change affect each other in intricate ways, from tiny plankton floating in the ocean to mammoth migrations that thunder across the African plains. This book focused on animal diversity, yet animal and plant diversity cannot be discussed in a completely separate manner. The plankton that fills the ocean builds the foundation for simple and complex animals; migrations of grazing animals help give new life to grasslands. By studying the nature of biodiversity, we learn about the Earth as a single, massive ecosystem.

Human civilization and its advances in the form of vehicles, construction, and mass production have caused a dreadful combination of pollution, animal extinctions, depletion of natural resources, and global warming. Humans can be proud of their ingenuity in creating products that enrich lives, but many of society's most important innovations have also consumed large amounts of resources. Today's generation and the next few generations must realize that the world's natural resources are not limitless. We have begun to reach the last reserves of many natural resources. In the case of thousands of animal species, the end arrived some time ago. Animals are going extinct at a rate never before seen in Earth's history, and a large part of this calamity can be traced back to the technological advances of human civilization.

The next generation of scientists will be called on to devise innovations not just for human comforts but to save animal species that have paid dearly for human inventions in the form of habitat loss. Making matters more complex, many habitats are doing no better than the animals. Impoverished regions contain people who are at this moment trying to carve out an existence on land tainted by pollution, dried up by climate change, and denuded of the natural resources that would give them a chance for a healthy life. In desperation, many have further destroyed animal habitat or hunted animals that are already teetering toward extinction. The future of animal diversity is therefore connected to the health and well-being of human society. Societies threatened by hunger, drought, poverty, or war become indirect and direct threats to endangered animals. Discussions on biodiversity lead beyond environmental science to sociology, ethics, politics, religion, economics, and human history.

To protect biodiversity, humans must make decisions on the value of life other than human life. Most progressive industrialized societies know that it is critical to preserve biodiversity. Without it, ecosystems falter and the overall health of the planet, and of human well-being, declines. But putting a value on nonhuman life creates a challenge that even the best minds have never overcome. It may still be easier for science and technology to save biodiversity than to value it.

Biodiversity experts confront their own thorny problems during their studies. Foremost among these is the undertaking of measuring three related indexes: the number of species on Earth, the number of extinctions taking place, and a real number calculated to define biodiversity. The number of species on Earth is estimated at about 14 million, but this could be an underestimate by as much as hundredfold or at least tenfold. Some species develop, exist, and go extinct without science ever having recorded them. Still, biologists know that the rates of animal and plant extinctions have increased since human civilization began, and these rates have accelerated since the onset of industrialization. Biologists have devised ways to put a number value on biodiversity, with the understanding that the value is only as good as the information science currently holds on animal species. Parameters such as richness, rarity, and rarity-weighted richness index—a way to draw a connection between richness and rarity—help put a number on biodiversity, but scientists and the public assume that the problem is far more complicated. The biodiversity parameters nonetheless help identify places with the richest biodiversity and those where animal losses have reached crisis levels.

With these numbers in hand, what do scientists as well as nonscientists now do about biodiversity loss? Somehow environmentalists must attract a critical mass of concerned individuals who understand the gravity of biodiversity loss. This requires a meeting of the minds from various sectors: the public, national and local governments, international education and watchdog groups, economists, religious leaders, and of course environmental scientists. Clearly, biodiversity is in serious trouble if its health depends on building a consensus among all of these groups. The near future in biodiversity science requires consensus building among diverse viewpoints so that real progress can be made in saving endangered species.

Naturalists have long pondered the worth of animal and plant life, as well as the nonliving things that make up their habitats. They developed the idea that everything in nature holds either instrumental or intrinsic value to humans. Perhaps the key to rousing human societies to stop destroying the Earth is putting a tangible price tag on nature. As crude as this approach may sound to those who love nature for itself, turning biodiversity into a business-related resource may be the only way to save it.

Technologies that aid endangered animals will improve without a doubt. Cooperation among human ideologies comes more slowly, but at certain pivotal points in history people have actually learned to work together, if temporarily. Biodiversity needs this type of global cooperation, and most important, critically endangered species need this help immediately. It is difficult to say whether we will find a way to stop the environmental destruction that takes food, land, and health away from wildlife. Those who volunteer or work in disciplines trying to save the environment hope for an answer soon. In the interim, a dedicated few represent the only chance of some endangered species to avoid premature extinction. Biodiversity is in peril, but hope always exists, and science has through the ages come forward to encourage that hope for things such as curing disease or changing the way the world communicates. Biodiversity is without argument the next critical challenge.

Appendix A

ECOLOGICAL FOOTPRINTS OF NATIONS, 2006 (ACRES/PERSON[1])			
NATION	ECOLOGICAL FOOTPRINT	CARBON FOOTPRINT	ECOLOGICAL DEFICIT (–) OR RESERVE (+)
Afghanistan	0.3	0.03	+0.5
Argentina	5.6	1.71	+8.9
Australia	16.2	8.42	+14.5
Austria	12.2	6.96	-3.8
Bangladesh	1.3	0.22	-0.5
Brazil	5.3	0.92	+19.2
Canada	18.8	10.08	+17.0
Chile	5.8	1.48	+7.5
China	4.1	1.86	-2.1
Costa Rica	4.9	1.59	-1.2
Croatia	7.3	4.12	-0.9
Denmark	14.2	7.84	-5.5
Egypt	3.3	1.26	-2.2
El Salvador	3.4	1.14	-2.0
Ethiopia	2.0	0.12	-0.7

(continues)

ECOLOGICAL FOOTPRINTS OF NATIONS, 2006
(ACRES/PERSON[1])   *(continued)*

NATION	ECOLOGICAL FOOTPRINT	CARBON FOOTPRINT	ECOLOGICAL DEFICIT (–) OR RESERVE (+)
France	13.9	5.00	-6.5
Germany	11.2	6.05	-6.9
Greece	12.4	7.84	-8.8
Guatemala	3.2	0.99	0.0
Guinea	2.3	0.14	+4.5
Honduras	3.1	1.02	+1.2
India	1.9	0.63	-0.9
Iran	5.9	3.76	-3.9
Ireland	12.2	7.70	-0.4
Israel	11.4	7.11	-10.5
Italy	10.3	6.22	-7.8
Japan	10.8	6.04	-8.9
Jordan	4.4	2.04	-3.7
Kazakhstan	9.8	6.72	+0.3
Kenya	2.0	0.37	-0.4
Korea, Republic of	10.0	4.84	-8.7
Liberia	1.7	0.02	+6.0
Libya	8.5	6.26	-6.0
Mexico	6.3	2.91	-2.2
Myanmar	2.3	0.11	+1.0
Nepal	1.7	0.23	-0.5

Nation	Ecological Footprint	Carbon Footprint	Ecological Deficit (–) or Reserve (+)
Nigeria	2.9	0.56	-0.6
Pakistan	1.5	0.51	-0.6
Peru	2.1	0.0	+7.3
Romania	5.8	2.59	-0.2
Russia	10.9	6.53	+6.1
Saudi Arabia	11.5	8.46	-9.1
Somalia	1.0	0.0	+0.8
South Africa	5.7	3.33	-0.6
Sudan	2.5	0.26	+1.9
Sweden	15.0	2.62	+8.7
Switzerland	12.7	6.85	-8.9
Tanzania	1.7	0.13	+1.4
Turkey	5.1	2.31	-1.7
Uganda	2.7	0.12	-0.6
United Arab Emirates	29.3	22.37	-27.2
United Kingdom	13.8	7.93	-9.8
United States	23.7	13.97	-12.0
Venezuela	5.4	2.84	+0.5
Yemen	2.1	0.77	-1.1
Zambia	1.5	0.22	+6.9
Zimbabwe	2.1	0.53	-1.1
WORLD	5.5	2.64	-1.1

Source: Global Footprint Network

[1] Convert acres to km^2 by multiplying by 0.004.

Appendix B

INTERNATIONAL ANIMAL CONSERVATION ORGANIZATIONS		
ORGANIZATION	**HEADQUARTERS**	**WEB SITE**
Conservation International	Arlington, Va.	www.conservation.org
Convention on International Trade in Endangered Species of Wild Fauna and Flora (CITES)	Geneva, Switzerland	www.cites.org
Convention on Migratory Species (Bonn Convention)	Bonn, Germany	www.cms.int
Defenders of Wildlife	Washington, D.C.	www.defenders.org
Greenpeace International	Amsterdam, The Netherlands	www.greenpeace.org/international
National Wildlife Federation	Reston, Va.	www.nwf.org
Nature Conservancy	Arlington, Va.	www.nature.org
Ocean Conservancy	Washington, D.C.	www.oceanconservancy.org
United Nations Environment Programme	Nairobi, Kenya	www.unep.org
United States Fish and Wildlife Service	Washington, D.C.	www.fws.gov

Organization	Headquarters	Web site
Wildlife Trade Monitoring Network (TRAFFIC)	Cambridge, United Kingdom	www.traffic.org
World Conservation Monitoring Center	Cambridge, United Kingdom	www.unep-wcmc.org
World Conservation Union	Gland, Switzerland	http://cms.iucn.org

Note: All Web sites are active as of March 1, 2009.

Appendix C

GLOBAL DIVERSITY HOTSPOTS		
NAME OF HOTSPOT	GENERAL REGION	PERCENT REMAINING, 2007
Atlantic forest	Brazil coast and parts of Paraguay, Argentina, and Uruguay	8
California floristic province	California and southwest Oregon coast	25
Cape floristic region	coastal tip of South Africa	20
Caribbean islands	all islands in the Caribbean Sea	10
Caucasus	parts of Russia, Georgia, Azerbaijan, Armenia, and Turkey	27
Central Asia mountains	parts of China, Kazakhstan, Kyrgyzstan, Tajikistan, Uzbekistan, and Turkmenistan	20
Cerrado	central Brazil	22
Chilean rainfall-Valdivian forests	Pacific coast of Chile	30
Coastal forests of eastern Africa	East Africa coast	10

Name of Hotspot	General Region	Percent Remaining, 2007
East Melanesian Islands	Solomon Islands and Vanuatu in the Pacific Ocean	30
Eastern Afromontane	parts of Congo, Uganda, Tanzania, and Zambia	11
Guinean forests, West Africa	southern coast of western Africa	15
Himalaya	Himalayan range from Pakistan through Myanmar	25
Horn of Africa	eastern tip of Africa and coasts of Arabian Peninsula	5
Indo-Burma	eastern India through Southeast Asia	5
Irano-Anatolian	parts of Iran, Iraq, Turkey, and Armenia	15
Japan	Japan islands	20
Madagascar, Indian Ocean islands	Madagascar and nearby islands	10
Madrean pine-oak woodlands	regions in central Mexico	20
Maputaland-Pondoland-Albany	southeastern coast of Africa	25
Mediterranean basin	coasts and islands of Mediterranean Sea	5
Mesoamerica	Central America, Mexico to Panama	20

(continues)

GLOBAL DIVERSITY HOTSPOTS *(continued)*

NAME OF HOTSPOT	GENERAL REGION	PERCENT REMAINING, 2007
New Caledonia	New Caledonia and nearby islands in the Coral Sea	5
New Zealand	New Zealand and nearby islands in the Pacific Ocean	22
Philippines	Philippine archipelago	7
Polynesia-Micronesia	Micronesia islands, Polynesia, and Fiji	21
Southwest Australia	southwestern tip of Australia	30
Southwest China Mountains	central southeastern China	8
Succulent Karoo	southwestern coast of Africa	29
Sundaland	Western Indonesia, Malaysia, Brunei, and Singapore	7
Tropical Andes	eastern South America, Venezuela to Argentina	16
Tumbes-Choco-Magdalena	southern coast of West Africa	24
Wallacea	central and eastern Indonesia	15
Western Ghats, Sri Lanka	western coast of India and Sri Lanka	23

Source: Conservation International

Glossary

ADAPTATION   acquiring traits from parents that help an individual or a species survive.

ADAPTIVE RADIATION   long-term process in which new species develop to fill new niches.

ADAPTIVE TRAIT   any inherited trait that enables an animal to survive better in a changed environment.

AFFLUENZA   unsustainable consumption of products and natural resources.

BIOCAPACITY   Earth's capacity to support its biota by providing resources necessary for life.

BIOCULTURAL RESTORATION   restoring a habitat to benefit animals and a local community at the same time.

BIODIVERSITY   variety of different species, genes within a species, or different ecosystems.

BIODIVERSITY HOTSPOTS   areas of greatest diversity and also greatest potential for habitat loss.

BIODIVERSITY INDICATOR   species that serves as an early warning that an ecosystem or a habitat is being damaged.

BIODIVERSITY INTACTNESS INDEX (BII)   percentage score calculated to show the portion of a population that remains compared with the original population before human activities began.

BIOGEOCHEMICAL CYCLE   natural recycling of Earth's nutrients in various chemical forms between living and nonliving things.

BIOGEOGRAPHIC REGION   large terrestrial area defined by climate and topography.

BIOLOGICAL COMMUNITY   populations of different species all living in an area and interacting with one another.

BIOME   a terrestrial area defined by the things living there, especially vegetation.

BIOREGION   land or water area defined by the communities of organisms living there and environmental characteristics.

BIOTA   living things; plant, animal, and microbial.

BUFFERING EFFECT   ability of an ecosystem to withstand damaging changes in the environment.

CARBON FOOTPRINT   a calculation of the total amount of greenhouse gases produced from a person or population's activities minus activities that reduce greenhouse gas outputs.

CARRYING CAPACITY   maximum population of a species that a habitat can support in a given period of time.

CHANGE DATA   information gathered on a geographic region that defines certain changes over time in the region.

CLUMPED DISTRIBUTION   species distribution in which members are part of groups, such as herds or flocks.

COMMENSALISM   interaction between two species in which one benefits and the other is neither helped nor harmed.

COMPETITIVE EXCLUSION   situation in which a species dominates resources in a habitat or ecosystem, forcing other species to leave in search of new resources.

COMPETITOR   two members of the same species or members of different species trying to use the same limited resources.

CONSERVATION   prudent use of the world's natural resources so that the resources are not used up rapidly or at all.

CONSUMERS   animal species that depend on getting all or a portion of their nutrients by eating plants or other consumers.

DECOMPOSERS   organisms that break down dead and decaying material for their metabolism and in the process help recycle the Earth's nutrients.

DELISTING   removing an animal from the U.S. Fish and Wildlife Service's endangered species list.

DOMINANCE   species distribution in which biologists see more individuals of a species than expected.

DOWNLISTING   moving a species from a highly threatened to a less threatened category on an endangered species list.

ECOLOGICAL DIVERSITY   variety of biological communities that interact with one another and with nonliving things in the environment.

ECOLOGICAL FOOTPRINT   amount of land and water needed to provide a population with natural resources to sustain life and dispose of wastes.

ECOLOGICAL NICHE the role of a species in an ecosystem and all the ecosystem factors that enable the species to live and reproduce.

ECOLOGICAL SUCCESSION process in which plant or animal species are replaced by other species over time, usually progressing from simple to complex organisms.

ECOSYSTEM community of species interacting with one another and with the nonliving things in a certain area.

ECOSYSTEM DIVERSITY variety of ecosystems in biology or the variations within an ecosystem.

ECOTOURISM traveling to areas for the purpose of observing unfamiliar animals, plants, or landscapes.

EDGE EFFECT changes in animal behavior or feeding patterns at places where two ecosystems meet.

ENDANGERED SPECIES wildlife or plants with so few individuals in their population they could soon become extinct in their natural range.

ENDEMISM existence of a species in only one area.

ENERGY PYRAMID diagram depicting the flow of energy through each level of a food chain.

EUTROPHICATION situation in which water pollution causes rapid growth of microorganisms that deplete the water of oxygen.

EVENNESS characteristic of a species distribution in which the numbers of individuals are similar in different areas over a certain time period.

EXOTIC SPECIES any nonnative plant or animal that has come from a very different habitat and is rare or highly valued.

EXPLOITATIVE COMPETITOR competitor that can remove an entire species from a food chain.

EXTINCTION complete disappearance of a species from Earth.

EXTINCTION SPASM event occurring every few to several centuries in which very large numbers of species go extinct within the same time period.

FITNESS ability of a species to thrive in an ecosystem or habitat.

FLAGSHIP SPECIES species monitored by humans to represent many species and their condition in a threatened environment, often used as a symbol of a specific habitat.

FOUNDATION SPECIES species that creates or enhances habitat that benefits other species.

FREE-ACCESS RESOURCES resources that are free and not owned or controlled by any person or jurisdiction.

FUNCTIONAL DIVERSITY   variety of biological and chemical processes that make energy and nutrients available for living things.

FUNCTIONAL GROUP   more than one species that share a similar characteristic, such as insect-eating birds and lizards.

FUNDAMENTAL NICHE   full range of resources a species could use if it had no competition.

GENE POOL   collection of all the genes in a particular population of individuals.

GENERALIST   species that can live in different habitats, eat a variety of foods, and tolerate a wide range of environmental conditions.

GENETIC DIVERSITY   variety in the genetic makeup of a species or a population within a species.

GENOME   complete set of genetic information in an organism.

GLOBAL WARMING   ongoing rise in temperatures at Earth's surface due to the accumulation of greenhouse gases in the atmosphere.

GROSS PRIMARY PRODUCTIVITY   rate at which producers capture and store all the Sun's available energy.

HABITAT   place where a plant or animal lives.

HABITAT ACCOMMODATION MODEL   theory proposing that animals enter an area undergoing succession only when the vegetation there can support them, and as plant succession progresses, the animal population changes to make best use of the vegetation.

HOTSPOT   any region known to have a rich content of biodiversity but also under great threat of being destroyed.

INDICATOR   any species that serves as an early warning of environmental damage.

INSTRUMENTAL VALUE   value of an organism, species, ecosystem, or biodiversity for the benefits it offers people.

INTERFERENCE COMPETITOR   invasive species that produces toxins harmful to ecosystems.

INTERSPECIFIC COMPETITION   situation in which two or more species compete for the same limited resources in a habitat or ecosystem.

INTRASPECIFIC COMPETITION   situation in which two or more members of the same species compete for the same limited resources in a habitat or ecosystem.

INTRINSIC VALUE   value of an organism, species, ecosystem, or biodiversity for itself and not because it may benefit people.

INTRODUCED SPECIES any nonnative plant or animal that has been deliberately put into an environment.

INVASIVE SPECIES species not normally found in a habitat or ecosystem but intentionally or accidentally released there.

KEYSTONE SPECIES species on which other species in an ecosystem depend.

MICROENVIRONMENT unique and usually small environment inhabited only by species possessing special features that enable them to live there.

MICROEVOLUTION small genetic changes in a population that allow it to react to changes in the environment.

MIGRATION movement of a species' population to a new region due to genetic factors, competition from other species, or natural events such as flood or drought.

MIGRATION CORRIDOR area of land, water, or air through which migrating populations pass.

MONOCULTURE growth of all the same one or two species of plants or trees in an area.

MUTUALISM interaction between two species in which both benefit.

NATURAL CAPITAL Earth's natural resources—chemical, physical, and biological process—that sustain life.

NET PRIMARY PRODUCTIVITY rate at which producers make energy available to consumers.

NICHE a species' role in an ecosystem.

NICHE DIFFERENTIATION (*also* niche splitting) a way two species reduce competition between themselves by using a habitat's resources differently.

NICHE SPECIALIZATION development by two species of specialties that allow them to occupy separate niches in the same ecosystem.

ONE-WAY MIGRATION permanent relocation of a population from one area to another area, usually due to loss of habitat or competitive exclusion.

OPPORTUNIST species able to take advantage of new conditions in the environment or to overcome drastic reductions in population due to changes in the environment.

PARASITISM interaction between two species in which one organism, the parasite, preys on another organism, the host, and causes harm to the host.

PIONEER SPECIES   first species that colonizes a site or a species that enters a habitat for the first time.

PLANKTON   tiny plants, animals, and microorganisms in aquatic environments that serve as a foundation for food chains.

POACHING   illegal killing, injuring, or removal from the habitat of endangered or threatened species.

POPULATION   group of individuals of the same species living in the same general area.

POPULATION CRASH   sudden decrease in the size of a population.

POPULATION DYNAMICS   major living and nonliving factors that cause an increase, decrease, or change in the composition of a species.

PREDATOR SPECIES   species that captures and feeds on other species.

PREMATURE EXTINCTION   disappearance of a species at a rate much faster than would naturally occur.

PRESERVATION   setting resources aside to be left unspoiled and undisturbed.

PRODUCERS   organisms that uses the Sun's energy (plants) or chemical energy (microorganisms) to make complex organic compounds from simple inorganic compounds, usually by photosynthesis.

RARITY   measure of the infrequency of a species living in an area.

RARITY-WEIGHTED RICHNESS INDEX (RWRI)   relationship between richness, or ample biodiversity, and rarity, or species infrequency; used for assessing biodiversity while accounting for species that are naturally rare.

REALIZED NICHE   the parts of a niche actually used by a species.

RECONCILIATION ECOLOGY   the study of restoring a habitat so it can be used by both wildlife and humans.

RED LIST   listing of all known animal and plant species categorized by status ranging from critically endangered to abundant.

REINTRODUCTION   release of an animal species into its natural habitat in a way that assures it will repopulate the habitat.

RESOURCE PARTITIONING   two species dividing up the resources in an ecosystem by using them at different times or in different places.

RESTORATION ECOLOGY   the study of repairing and reconstructing damaged habitat.

RICHNESS   measure of the number and frequency of species living in an area; measure of diversity.

RIGS-TO-REEFS   process of leaving abandoned offshore oil rigs in place to serve as habitat for various aquatic species.

RIPARIAN   describing area along stream banks, riverbanks, marshes, or shorelines.

SPECIATION   formation of two species from one because of divergent adaptations and natural selection.

SPECIES   group of organisms that resemble each other in genetic makeup, appearance, and behavior and that breed with each other.

SPECIES DIVERSITY   variety of species and their relative abundance in an area.

THREATENED SPECIES   wildlife or plants abundant in their natural range but likely to become endangered because of declining numbers.

TRANSGENIC   organism containing one or more genes from another organism.

TROPHIC CASCADE   change in one species' population leading to major disturbances of other, unrelated species.

WEB OF LIFE   complex network of many interconnected food chains.

Further Resources

PRINT AND INTERNET

Austin, Jay E., and Carl E. Bruch, eds. *The Environmental Consequences of War.* Cambridge: Cambridge University Press, 2000. Available online. URL: http://books.google.com/books?id=OJKFkSkgTyIC. Accessed March 1, 2009. The authors take a somber look at one aspect of biodiversity loss.

The Australian. "Ecological Mapping to Protect Antarctic." November 1, 2006. Available online. URL: www.theaustralian.news.com.au/story/ 0,20867,20682898-30417,00.html. Accessed March 1, 2009. A resource that explains a technology used in biodiversity study.

Barringer, Felicity. "Polar Bear Is Made a Protected Species." *New York Times,* 15 May 2008. Available online. URL: www.nytimes.com/2008/05/15/us/ 15polar.html?_r=2&scp=3&sq=polar+bear&st=nyt&oref=slogin&oref=slog in. Accessed March 1, 2009. This article provides an update on endangered species decisions by the federal government.

Beisker, Greg. "John F. Lacey: Champion for Birds and Wildlife—Iowa's (Almost) Forgotten Conservationist." Iowa Natural Heritage Foundation. Available online. URL: www.inhf.org/lacey/overview2.htm. Accessed March 1, 2009. An online article providing details on the history of the Endangered Species Act.

Benjamin, Alison. "Threatened Species Red List Shows Escalating 'Global Extinction Crisis'." United Kingdom *Guardian,* 12 September 2007. Available online. URL: www.guardian.co.uk/environment/2007/sep/12/international-news.greenpolitics. Accessed March 1, 2009. The current status of endangered species with opinions from world biodiversity experts.

Berke, Richard L. "Clinton Declares New U.S. Policies for Environment." *New York Times* (4/22/93). Available online. URL: http://query.nytimes.com/gst/ fullpage.html?res=9F0CEEDB1F38F931A15757C0A965958260&scp=5&sq= biodiversity+treaty&st=nyt. Accessed March 1, 2009. A recap on the content of President Bill Clinton's first major speech of his administration on the environment.

Blignaut, James, and James Aronson. "Getting Serious about Maintaining Biodiversity." *Conservation Letters* 1 (2008): 12–17. General background and updates on conservation.

Borenstein, Seth. "Narwhals on Thinner Ice Than Polar Bears." *San Francisco Chronicle* (4/26/08). A discussion of Arctic species less well-known than polar bears that are also in habitat crisis.

Breed, Allen G. "Rare Woodpecker Has Town Felling Trees." *San Francisco Chronicle* (9/24/06). A description of unintended and damaging consequences of the endangered species list.

Brown, Matthew. "Endangered Wolves Cause Danger of Their Own." *San Francisco Chronicle* (11/23/07). Explains the debate on removing the gray wolf from the endangered species list.

Brown, Matthew. "Gray Wolves of the West No Longer 'Endangered'." *San Francisco Chronicle* (2/22/08). Coverage of the heated arguments regarding the delisting of gray wolves from the endangered species list.

Catlin, George. *Letters and Notes on the Manners, Customs, and Conditions of the North American Indians.* 1841. Whitefish, Mont.: Kessinger, 2005. Available online. URL: http://books.google.com/books?id=opn6KyHfdVwC. Accessed March 1, 2009. A compilation providing an interesting view of an American naturalist.

Chapman, Jenny L., and Michael J. Reiss. *Ecology: Principles and Applications,* 2nd ed. New York: Cambridge University Press, 1999. A well-illustrated book with basic principles of ecology related to all aspects of biodiversity.

Christensen, Jon. "Interloper Ants Keep It All in the Family." *New York Times,* 1 August 2000. Available online. URL: http://query.nytimes.com/gst/fullpage.html?res=9503E2D6133DF932A3575BC0A9669C8B63&sec=&spon=&pagewanted=2. Accessed March 1, 2009. A description of the status of invasive Argentine ants in North America.

Clodfelter, Lindsey. "Island Pig Eradication Spurs Wild Controversy." University of California, Santa Barbara, *Daily Nexus,* 2 March 2005. Available online. URL: www.dailynexus.com/article.php?a=9160. Accessed March 1, 2009. A description of the controversies surrounding a local invasive species eradication program.

Coalition for Enhanced Marine Resources. "What's New with CEMR and Rigs-to-Reefs." *Reef Source* 3 (2006): 1–3. Options for creating artificial habitat out of retired oil-drilling platforms.

Coleman, Joseph. "Electricity Helps Restore Coral Near Bali." *San Francisco Chronicle,* 6 December 2007: A22. A new technology for returning health to damaged coral.

Collis, Brad. "The Beetle That Saved Lake Victoria." *Australian Broadcasting Corporation*. 2000. Available online. URL: www.abc.net.au/science/slab/hyacinth/default.htm. Accessed March 1, 2009. A television report explains the Lake Victoria restoration project.

Darwin, Charles. *On Origin of Species*. 1859. Reprinted with foreword by George G. Simpson. New York: Collier Books, 1962. The classic text in evolution studies.

———. *The Voyage of the Beagle*. 1839. First published as *Journal of Researches*. New York: P.F. Collier and Sons, 1909. Available online. URL: http://books.google.com/books?id=MDILAAAAIAAJ&dq=the_voyage+of+the+beagle&source=gbs_summary_s&cad=0. Accessed March 1, 2009. Darwin's classic journal of nature observations on Galápagos and several other Pacific islands.

Duplancic, Neno. "Locus Technologies Seeks to Pioneer Environmental Information Management through On-Demand Approaches." Interview with *Environmental Business Journal* 21, no. 1/2 (2008): 33–35. Company president explains new techniques in environmental imaging.

Durnan, Kimberly. "PETA Asks Dallas Zoo to Close Elephant Exhibit." *Dallas Morning News,* 16 May 2008. Available online. URL: www.dallasnews.com/sharedcontent/dws/news/localnews/stories/051708dnmetelephant.107b8c007.html. Accessed March 1, 2009. Background on the role of wildlife sanctuaries.

Eldredge, Niles. *Life in the Balance: Humanity and the Biodiversity Crisis*. Princeton, N.J.: Princeton University Press, 1998. A thought-provoking book on biodiversity.

European Commission and the World Conservation Union. *Biodiversity in Development: The Links between Biodiversity and Poverty*. Available online. URL: www.undp.org/biodiversity/biodiversity/BioBrief1-poverty.pdf. Accessed February 28, 2009. A clear explanation of the biodiversity-poverty connection.

Fogarty, Michael J., and Steven A. Murawski. "Do Marine Protected Areas Really Work?" *Oceanus* (5/24/08). Available online. URL: www.whoi.edu/oceanus/viewArticle.do?id=3782&archives=true&sortBy=printed. Accessed March 1, 2009. An interesting background on protected marine areas by focusing on Georges Bank.

Gaston, Kevin J., and John I. Spicer. *Biodiversity: An Introduction,* 2nd ed. Oxford: Blackwell, 2004. A clearly written textbook that delivers its message using current examples from around the world.

Gore, Al. *An Inconvenient Truth*. New York: Rodale, 2006. Vice President Gore surveys the world's current environmental crisis.

Hagengruber, James. "Crowd Favors Getting Rid of Wolves—Idaho Anti-Wolf Coalition Says Support Growing." *Idaho Spokesman-Review* (1/12/04). Available online. URL: www.citizenreviewonline.org/jan2004/crowd.htm. Accessed March 1, 2009. A description of a public demonstration regarding gray wolves in the Rocky Mountains.

Harmer, Jerry. "Ho Chi Minh Trail Area Safe for Wildlife." *Boston Globe* (5/3/07). Available online. URL: www.boston.com/news/world/asia/articles/2007/03/03/ho_chi_minh_trail_area_safe_for_wildlife/?page=1. Accessed March 1, 2009. An interesting glimpse at Vietnam's tiger-monitoring program.

Herbert, Ian, and Jonathan Brown. "Your Carbon Footprint Revealed: Climate Change Report Finds We Each Produce 11 Tons of Carbon a Year—And Breaks Down How We Do It." London *Independent,* 9 December 2006. An explanation of carbon footprint scores.

Intergovernmental Panel on Climate Change. *Climate Change 2007: Synthesis Report.* Available online. URL: www.ipcc.ch/pdf/assessment-report/ar4/syr/ar4_syr.pdf. Accessed March 1, 2009. A lengthy agency report providing an excellent resource on climate change with very good illustrations.

International Herald Tribune. "A 1970 Plan for a Tire Reef off Florida Turns into an Ecological Disaster" (2/18/07). Available online. URL: www.iht.com/articles/2007/02/18/news/tires.php?page=1. Accessed March 1, 2009. The background on Florida's experiment to replace dead coral reefs with artificial tire-reefs.

Iredale, Will. "Polar Bears Drown as Ice Shelf Melts." London *Sunday Times* (12/18/05). Available online. URL: www.timesonline.co.uk/tol/news/uk/article767459.ece. Accessed March 1, 2009. One of the earliest reports of habitat loss suffered by polar bears due to ocean warming.

Kay, Jane. "Ocean Warming's Effect on Phytoplankton." *San Francisco Chronicle* (12/7/06). Describes why warming ocean temperatures disrupt ocean ecosystems.

———. "Grim Global Warming Prognosis for Western U.S." *San Francisco Chronicle* (2/3/07). The effect of global warming on drought.

KQED-TV. "San Francisco Bay Invaders." Aired May 29, 2007. Available online. URL: www.kqed.org/quest/television/view/332. Accessed March 1, 2009. A half-hour broadcast on how invasive species affect ecosystems.

Leopold, Aldo. *Game Management.* Madison: University of Wisconsin Press, 1986. A classic text on conservation still used in ecology today.

Leopold, Aldo, Luna B. Leopold, and Charles W. Schwartz. *Round River: From the Journals of Aldo Leopold.* Oxford: Oxford University Press, 1972. An insightful collection from Leopold's journals and essays.

Lester, Greg. "Genetic Therapy Reverses Nervous System Damage." University of Pennsylvania School of Veterinary Medicine *Bellwether,* Summer 2005. A resource to explain gene therapy.

Lewis, Meriweather. *Journal of Lewis and Clark,* 1804–1806. Available online. URL: www.lewisandclarktrail.com/diary.htm. Accessed March 1, 2009. The original text from the Lewis and Clark expedition provides fascinating insight into their discoveries and Lewis's description of natural things.

Lovejoy, Thomas E., and Lee Hannah, eds. *Climate Change and Biodiversity.* New Haven, Conn.: Yale University Press, 2004. A standard resource.

McGrath, Susan. "Attack of the Alien Invaders." *National Geographic,* March 2005. A well-illustrated article on alien species, especially plants.

Mittermeier, Russell A., Patricio R. Gil, Michael Hoffman, John Pilgrim, Thomas Brooks, Cristina G. Mittermeier, John Lamoreux, and Gustavo A. B. da Fonseca. *Hotspots Revisited.* Arlington, Va.: Conservation International, 2005. Offers striking images of the world's hotspots with thought-provoking text.

Moir, John. "The Face of Recovery." *Birder's World,* December 2007. Available online. URL: www.birdersworld.com/brd/default.aspx?c=a&id=998. Accessed March 1, 2009. An article describing the California condor recovery program.

Myers, Norman. "The Journey of an Environmental Scientist." Interview with Harry Kreisler. Conversations with History, University of California. March 1, 2009. Available online. URL: http://globetrotter.berkeley.edu/conversations/m.html. Accessed November 22, 2008. Interview with one of the foremost biodiversity experts.

Nature Conservancy. "Big Woods Habitat Protected through Land Deal between the Nature Conservancy and U.S. Fish and Wildlife Service." Press release. September 29, 2005. Provides information on the protection plans for the ivory-billed woodpecker.

———. "Palau: A Champion of Coral Reefs." Available online. URL: www.nature.org/wherewework/asiapacific/micronesia/features/mcpalau.html. Accessed March 1, 2009. This online article explains Palau's plans as a protector of biodiversity.

New York Times. "Gale Norton Resigns." (3/12/06). Available online. URL: www.nytimes.com/2006/03/12/opinion/12sun2.html. Accessed March 1, 2009. An editorial that reviews the environmental record of outgoing Secretary of the Interior Gale Norton.

Norris, S., L. Rosentrater, and P. M. Eid. *Polar Bears at Risk. A WWF Status Report.* Oslo: World Wildlife Fund, 2002. Available online. URL: www.wwf.org.uk/filelibrary/pdf/polar_bears_at_risk_report.pdf. Accessed March 1, 2009. A booklet that provides a detailed examination of endangered species and the effects of climate change.

Northern Prairie Wildlife Research Center, U.S. Geological Survey. "Migration of Birds." Available online. URL: www.npwrc.usgs.gov/resource/birds/migratio. Accessed March 1, 2009. An interesting review of the principles of migration.

Nyberg, Kara. "Reconciliation Ecology Could Save Species Headed for Extinction, UA Ecologist Says." University of Arizona *UA News,* 10 May 2001. Available online. URL: http://uanews.org/node/4818. Accessed March 1, 2009. This short piece describes the features of reconciliation ecology as a conservation tool.

Orange County Coastkeeper. "Rigs to Reefs: Hidden Oasis or Environmental Grief?" News release (3/30/07). Available online. Accessed February 28, 2009. Insight on the proposed plans to leave abandoned offshore oil rigs in place.

Oswell, Adam. "Black Market Animal Trade Threatens Endangered Species." Interview with Tanya Nolan. *ABC, The World Today,* August 1, 2005. Available online. URL: www.abc.net.au/worldtoday/content/2005/s1427351.htm. Accessed March 1, 2009. This article looks at a seven-year program to study the international illegal trade in endangered animals.

Pittman, Craig. "Everglades Restoration Bogs Down." *St. Petersburg Times* (7/3/07). Available online. URL: www.sptimes.com/2007/07/03/news_pf/State/Everglades_restoratio.shtml. Accessed March 1, 2009. An explanation of the conflicting forces that have hampered Florida's Everglades recovery project.

Rappold, R. Scott. "Gail Norton Defends Her Record." *Colorado Springs Gazette* (4/8/08). Available online. URL: www.gazette.com/articles/norton_35056___article.html/farmer_many.html. Accessed March 1, 2009. A short article covering retired Secretary of the Interior Gale Norton's views on the environment.

Revkin, Andrew C. "Biologists Sought a Treaty; Now They Fault It." *New York Times* (5/7/02). Available online. URL: http://query.nytimes.com/gst/fullpage.html?res=9F0DE0DF1730F934A35756C0A9649C8B63&sec=&spon=&pagewanted=3. Accessed March 1, 2009. An interesting article containing various opinions on the role of genetic engineering in biodiversity conservation.

———. "In a New Climate World, Short-Term Cooling in a Warmer World." *New York Times* (5/1/08). Available online. URL: www.nytimes.com/2008/05/01/science/earth/01climate.html?scp=1&sq_In+a+New+Climate+World%2C+Short-Term+Cooling+in+a+Warmer+World&st =nyt. Accessed March 1, 2009. This article describes how scientists measure climate change.

Ritter, John. "Calif. Hopes to Hook Lake's Pike Problem." *USA Today* (8/23/07). Available online. URL: www.usatoday.com/news/nation/environment/2007-08-23-lake-davis-trout_N.htm. Accessed March 1, 2009. Background on a nonnative fish eradication program in a lake ecosystem.

Roach, John. "Conservationists Name Nine New 'Biodiversity Hotspots'." National Geographic News (9/2/05). Available online. URL: http://news.national geographic.com/news/2005/02/0202_050202_hotspots.html. Accessed March 1, 2009. This article is dated but gives a clear explanation of the concept of biodiversity hotspots.

Roosevelt, Theodore. State of the Union Message, on December 3, 1907. UCSB American Presidency Project. Available online. URL: www.theodore-roosevelt. com/sotu7.html. Accessed March 1, 2009. This complete text of Roosevelt's speech provides insight into his thoughts on environment and conservation.

Science Daily. "Restoration of a Tropical Rain Forest Ecosystem Successful on Small-Scale" (4/30/08). Available online. URL: www.sciencedaily.com/releases/2008/04/080428133928.htm. Accessed March 1, 2009. Current conservation methods in Costa Rica are described.

Shanahan, Mike. "Measuring Loss of Biodiversity the Expert Way." *SciDev Net* (3/11/05). Available online. URL: www.scidev.net/en/news/measuring-loss-of-biodiversity-the-expert-way.html. Accessed March 1, 2009. Description of the biodiversity-intactness index.

Sibley, David A. *The Sibley Guide to Bird Life and Behavior*. New York: Chanticleer Press, Alfred A. Knopf, 2001. This handbook is a modern classic on the subject of bird ecology.

Simmons, Randy T. *Endangered Species*. Edited by Cynthia A. Bily. Farmington Hills, Mich.: Greenhaven Press, 2007. Essays with opposing viewpoints on the status of endangered species.

Skoloff, Brian. "Massive Everglades Restoration Project Halted." *San Francisco Examiner* (5/15/08). A history of the controversies of Florida's Everglades project.

Squatriglia, Chuck. "Desert Pupfish in Hot Water." *San Francisco Chronicle* (5/27/07). Interesting article about the rare species that initiated the Endangered Species Act.

Steffen, Alex, ed. *Worldchanging*. New York: Harry N. Abrams, 2006. Manual of green living and how to protect Earth from further environmental damage.

Stienstra, Tom. "Wildlife Detective Stars in a Real-Life Drama in a Forensic Lab." *San Francisco Chronicle* (12/17/06). Describes how DNA profiling of animals catches illegal hunters.

Tansley, Alfred G. "The Use and Abuse of Vegetational Concepts and Terms." *Ecology* 16, no. 3 (1935): 284–307. Journal article presenting the first academic description of the ecosystem concept.

Tran, Cathy. "A Platform for Debate." Orange County, Calif., *OC Register* (3/31/07). Available online. URL: www.ocregister.com/ocregister/homepage/abox/article_1637579.php. Accessed February 27, 2009. The debates surrounding oil-drilling platform rigs as artificial marine habitats.

United Nations Environment Programme, World Conservation Monitoring Centre. "Biodiversity indicators for national use." Available online. URL: www.unep-wcmc.org/collaborations/BINU. Accessed March 1, 2009. Excellent reference on indicator techniques and mapping.

Viegas, Jennifer. "On the Trail of a Great White." Discovery News (4/7/08). Available online. URL: http://dsc.discovery.com/news/2008/04/07/great-white-map.html. Accessed March 1, 2009. Description of the Monterey Bay Aquarium's project for breeding white sharks in captivity.

Waugh, Jeff. "Banff National Park—Wildlife: Poachers Threaten Wildlife." Available online. URL: www.canadianrockies.net/banff/poaching.html. Accessed March 1, 2009. Examination of the reasons behind wildlife poaching increases in Canada.

Weeks, Jennifer. "Wildlife Corridors—All Clear." *Audubon* (September–October 2007). Current projects and innovations in artificial wildlife corridors in North America.

The White House. President Clinton and Vice President Gore. "World's Largest Environmental Restoration Will Restore Natural Flow to Florida Everglades." December 11, 2000. Available online. URL: http://clinton4.nara.gov/WH/new/html/Mon_Dec_11_154136_2000.html. Accessed March 1, 2009. Online recounting of President Clinton and Vice President Gore's launch of the Everglades restoration project.

Wilderness Society. "Robert Marshall." Available online. URL: http://wilderness.org/about-us/bob-marshall. Accessed March 1, 2009. This biography recounts the career of the founder of the Wilderness Society.

Williams, Ted. "Back Off!" *Audubon* (May–June 2007). An opinion piece that gives a compelling account of the wolf-reintroduction controversy in the western United States.

Wilson, Edward O., ed. *Biodiversity.* Washington, D.C.: National Academies Press, 1988. The standard resource in academia on biodiversity and still useful though published more than 20 years ago.

———. *The Future of Life.* New York: Alfred A. Knopf, 2003. Wilson presents an entertaining discussion of species, the world's ecology, and Earth's future.

WEB SITES

African Conservation Centre. Available online. URL: www.conservationafrica.org. Accessed March 1, 2009. Description of current African wildlife conservation programs.

Association of Zoos and Aquariums. Available online. URL: www.aza.org. Accessed March 1, 2009. Descriptions of specialized conservation programs at zoos and aquariums.

Conservation International. Available online. URL: www.conservation.org/Pages/default.aspx. Accessed March 1, 2009. Excellent resource for international programs in terrestrial and aquatic biodiversity conservation.

Convention on Biological Diversity. Available online. URL: www.cbd.int. Accessed March 1, 2009. CBD provides information on international biodiversity programs.

Global Coral Reef Alliance. Available online. URL: www.globalcoral.org. Accessed March 1, 2009. Good resource on coral reefs and general marine biodiversity.

Global Footprint Network. Available online. URL: www.footprintnetwork.org/en/index/php/GFN. Accessed March 1, 2009. General resource on environmental issues pertaining to natural resources.

Nature Conservancy. Available online. URL: http://www.nature.org. Accessed March 1, 2009. General information on biodiversity; contains an online carbon footprint calculator.

United Nations Environment Programme. World Conservation Monitoring Center. Available online. URL: www.unep-wcmc.org. Accessed March 1, 2009. The main international resource on biodiversity.

U.S. Department of Agriculture, National Invasive Species Information Center. Available online. URL: www.invasivespeciesinfo.gov. Accessed March 1, 2009. USDA provides an introduction to invasive species from microbes to mammals.

U.S. Fish and Wildlife Service. Available online. URL: www.fws.gov. Accessed March 1, 2009. The U.S. agency serves as the home of the endangered species list.

Wildlife Preservation Canada. Available online. URL: www.wptc.org. Accessed March 1, 2009. Resource for conservation programs in Canada and other nations.

World Conservation Union. Available online. URL: www.iucn.org. Accessed March 1, 2009. Home of the IUCN Red List of Threatened Species.

World Resources Institute. Available online. URL: www.wri.org. Accessed March 1, 2009. Provides overviews of ecological issues related to population dynamics.

Yellowstone to Yukon Conservation Initiative. Available online. URL: http://www.y2y.net. Accessed March 1, 2009. Resource on details of how migration corridors are planned and maintained.

Index

Note: Page numbers in *italic* refer to illustrations. The letter *t* indicates tables.